RECENT DEVELOPMENTS IN SLIDING MODE CONTROL THEORY AND APPLICATIONS

Edited by **Andrzej Bartoszewicz**

Recent Developments in Sliding Mode Control Theory and Applications
http://dx.doi.org/10.5772/65568
Edited by Andrzej Bartoszewicz

Contributors

Sohom Chakrabarty, Bijnan Bandyopadhyay, Andrzej Bartoszewicz, Wen-Fang Xie, Shutong Li, Ahmad Ghasemi, Yanbin Gao, Zaouche Mohammed, Foughali Khaled, Amini Mohamed, Ahmadreza Argha, Steven Su, Zool H Ismail, Vina Wahyuni Eka Putranti

Published by InTech

Janeza Trdine 9, 51000 Rijeka, Croatia

Publishing Process Manager Marijana Francetic
Technical Editor SPi Global
Cover InTech Design team

First published June, 2017
Printed in Croatia
Legal deposit, Croatia: National and University Library in Zagreb

Additional hard copies can be obtained from orders@intechopen.com

Recent Developments in Sliding Mode Control Theory and Applications, Edited by Andrzej Bartoszewicz
p. cm.
Print ISBN 978-953-51-3271-4
Online ISBN 978-953-51-3272-1

For EU product safety concerns:
IN TECH d.o.o., Prolaz Marije Krucifikse Kozulić 3, 51000 Rijeka, Croatia,
info@intechopen.com or visit our website at intechopen.com.

Meet the editor

Andrzej Bartoszewicz received his MSc degree in 1987 and his PhD degree in 1993 both from Technical University of Łódź, Poland. He was visiting scholar at Purdue University, West Lafayette, IN, USA, and at Strathclyde University, Glasgow, UK. Then for 1 year, he was at the University of Leicester, UK. Currently, he is a professor at Technical University of Łódź, the head of Electric Drive and Industrial Automation Group, and the director of Institute of Automatic Control. He has published 3 monographs and over 300 papers, primarily in the field of sliding mode control. He received three best paper awards at international conferences. Professor Andrzej Bartoszewicz has recently been elected a corresponding member of the Polish Academy of Sciences.

Contents

Preface

The main purpose of control engineering is to steer the regulated plant in such a way that it operates in a required manner. The desirable performance of the plant should be obtained despite the unpredictable influence of the environment on all parts of the control system, including the plant itself, and no matter if the system designer knows precisely all the parameters of the plant. Even though the parameters may change with time, load, and external circumstances, still the system should preserve its nominal properties and ensure the required behavior of the plant. In other words, the principal objective of control engineering is to design systems that are robust with respect to external disturbances and modeling uncertainty. This objective may be very well achieved using the sliding mode technique, which is the subject of this book.

The theory of variable structure systems with sliding (or quasi-sliding) modes is currently one of the most significant research topics within the control engineering domain. Moreover, recently, a number of important applications of the theory have also been reported. Therefore, this book first presents two chapters concerning the theory of sliding mode control and then shows three significant applications of this control method.

In the first chapter, Argha and Su design a new, continuous time, suboptimal in the sense of $\mathcal{H}2$ and $\mathcal{H}\infty$ norm state-feedback sliding mode controller for output tracking of the closed-loop system. This approach enables to effectively trade off the level of control effort required to maintain sliding mode and the system performance. In the second chapter, Chakrabarty, Bandyopadhyay, and Bartoszewicz consider discrete time quasi-sliding mode control strategies with switching variables characterized by relative degree greater than one. They demonstrate that introduction of such variables can lead to desirable system performance and a better robustness than application of conventional, that is, relative degree one, switching variables.

The next three chapters are devoted to applications of sliding mode controllers in robotics and motion control systems. In the first of these chapters, Li, Ghasemi, Xie, and Gao present a sliding mode controller for image-based visual servoing of a six-degree-of-freedom robot arm. Their controller integrates the standard proportional-derivative action with the sliding mode approach to obtain faster convergence and increased robustness with respect to disturbances. The next chapter of this book is concerned with the important practical problem of autonomous underwater vehicle motion control. In this chapter, Putranti, Wahyuni and Ismail propose a novel controller that combines a popular super twisting sliding mode control paradigm with a region boundary approach. They demonstrate that defining the desired trajectory as a region can result in an energy saving when compared to conventional control with the demand state defined as a single point. In the final chapter of this book,

Zaouche, Foughali, and Amini consider adaptive integral higher order sliding mode control for an aircraft autopilot. Their control method exhibits good robustness and is simpler to implement than other standard techniques typically used in this application.

The contributions presented in this book demonstrate favorable properties of modern sliding mode control techniques, that is, their robustness and computational efficiency, and encourage new applications of these techniques in various engineering systems.

Andrzej Bartoszewicz
Institute of Automatic Control
Łódź University of Technology,
Łódź, Poland

Chapter 1

State-Feedback Output Tracking Via a Novel Optimal-Sliding Mode Control

Ahmadreza Argha and Steven W. Su

Additional information is available at the end of the chapter

http://dx.doi.org/10.5772/67468

Abstract

This chapter describes a new framework for the design of a novel suboptimal state-feedback-sliding mode control for output tracking while $\mathcal{H}_2/\mathcal{H}_\infty$ performances of the closed-loop system are under control. In contrast to most of the current sliding surface design schemes, in this new framework, the level of control effort required to maintain sliding is penalized. The proposed method for the design of optimal-sliding mode control is carried out in two stages. In the first stage, a state-feedback gain is derived using a linear matrix inequality (LMI)-based scheme that can assign a number of the closed-loop eigenvalues to a known value while satisfying performance specifications and ensuring that all the closed-loop poles are located in a preselected subregion. The sliding function matrix related to the particular state feedback derived in the first stage is obtained in the second stage by using one of the two different methods developed for this goal. We present a numerical example to demonstrate the remarkable performance of the proposed scheme.

Keywords: optimal $\mathcal{H}_2/\mathcal{H}_\infty$-based sliding mode control, output tracking, partial eigenstructure assignment, regional pole placement

1. Introduction

Sliding mode control (SMC) is now a well-developed method of control and its invariance properties against matched uncertainties have inspired researchers to apply this technique to different applications [1–6]. Traditionally, SMC is designed in two stages. In the first stage, a sliding surface whose sliding motions have suitable dynamics is chosen. Many methods have been proposed in the existing literature for this purpose, for example, eigenstructure assignment, pole placement, optimal quadratic methods, and linear matrix inequality (LMI) methods; see for instance [4, 5, 7, 8]. Then, a controller is designed to induce and maintain the sliding motion.

However, these traditional design methods are unable to limit the available control action required for satisfying the control objective [3], since during the switching-function synthesis, there is no sense of the level of the control action required to persuade and retain sliding [3]. It is worth noting that without having limits on the available control actions, a sliding surface and thereby a control law may always be obtained which is not practically applicable, as it may lead to high level of control efforts.

To tackle this problem, for instance, the authors of [9] propose a scheme to design a sliding surface which minimizes an objective function of the system state and control input, in the meantime. However, since the method in [9] needs to ensure that at least one eigenvalue of the closed-loop system (for single-input systems) is a real value, not necessarily any arbitrary weighting matrices in the objective function may result in a sliding mode control. This reference, therefore, either reselects the weighting matrices or approximates the closed-loop system eigenvalues so that a set of eigenvalues are generated which can be divided into the null-space and range-space dynamics. However, no precise scheme is given on how to reselect the weighting matrices. Further, the approximation of eigenvalues may lead to a loss in optimality and possibly robustness.

For addressing the limitations of [9], Tang and Misawab [10] propose an LQR-like scheme in which a weighting matrix is computed which is closest to the desired one and can result in the desired eigenvalues. Following this, the associated SMC is designed according to the obtained eigenvalues and weighting matrix. Nevertheless, both methods in [9, 10] are suitable to single-input systems. Alternatively, Edwards [3] proposes two new frameworks exploiting two special system coordinate transformations, which are fundamentally different from the aforementioned schemes.

This chapter aims to propose a different way for the sliding surface design while optimizing the control effort associated with the linear part of the control law. This approach is a middle-of-the-road method in that it uses a specific *partial eigenstructure assignment* method to assign m arbitrary stable real eigenvalues while an appropriate sliding motion dynamics will be ensured by enforcing different Lyapunov-type constraints such as the $\mathcal{H}_2/\mathcal{H}_\infty$ and regional pole-placement constraints. The advantages of the proposed approach for the design of sliding surface compared to all the aforementioned references are threefold: (i) it can set the stage for designing SMC while the level of control efforts is taken into account; (ii) it makes it possible to integrate several Lyapunov-type constraints, for example, regional pole-placement constraints, in the SMC design problem; and (iii) the controller can be computed in a numerically very efficient method. The proposed scheme for the design of suboptimal SMC is indeed a two-stage LMI-based approach. In the first stage, while enforcing different Lyapunov-type constraints, for example, the mixed $\mathcal{H}_2/\mathcal{H}_\infty$, a state-feedback gain is derived, using an LMI-based optimization program employing an instrumental matrix variable, which can precisely assign some of the closed-loop eigenvalues to a priori known value. Following this, the sliding surface, associated with the state-feedback gain obtained in the first stage, is determined in the second stage. Two different approaches are presented for deriving the associated switching-function matrix. This chapter indeed examines the problem of designing a state-feedback SMC which utilizes integral action to provide tracking. From the implementation point of view, the simplicity of such a scheme is very advantageous.

The structure of the chapter is as follows. Section 2 is dedicated to the problem statement and preliminaries. Section 3 explains the novel design strategy for the design of $\mathcal{H}_2/\mathcal{H}_\infty$-based SMC. Section 4 discusses two different approaches for deriving the sliding function matrix associated with the linear controller obtained in Section 3. Section 5 summarizes the proposed $\mathcal{H}_2/\mathcal{H}_\infty$-based SMC. In Section 6, we discussed the issue of designing SMC with additional regional pole-placement constraints. Section 7 illustrates this method via an example considering the flight control problem. Section 8 finally concludes the chapter.

2. Problem statement and preliminaries

Consider the following linear-time invariant (LTI) system:

$$\dot{\tilde{x}}(t) = \tilde{A}\tilde{x}(t) + \tilde{B}[u(t) + f(t)], \tag{1}$$

where $\tilde{x} \in \mathbb{R}^{\tilde{n}}$ and $u \in \mathbb{R}^m$ are the state vector and control input vector, respectively. The matrices in Eq. (1) are constant and of appropriate dimensions. The unknown signal $f(t) \in \mathbb{R}^m$ denotes matched uncertainty in Eq. (1) whose Euclidean norm is bounded by a known function $\rho(t)$. Without the loss of generality, it is assumed that rank $(\tilde{B}) = m$ and the matrix pair $(\tilde{A}, \tilde{B})$ are controllable.

In order to provide the problem with a tracking facility, we exploit an integral action as follows. Defining

$$\dot{\xi}(t) = \mathbb{r}(t) - \tilde{y}(t), \tag{2}$$

where $\mathbb{r}(t)$ is the input reference to be tracked by $\tilde{y}(t) = \tilde{C}\tilde{x}(t) \in \mathbb{R}^p$, and ξ represents the integral of the tracking error, that is, $\mathbb{r}(t) - \tilde{y}(t)$, and introducing $x := \begin{bmatrix} \xi \\ \tilde{x} \end{bmatrix} \in \mathbb{R}^n$, an augmented system can be derived as

$$\dot{x}(t) = Ax(t) + B_2u(t) + B_{\mathbb{r}}\mathbb{r}(t), \tag{3}$$

with

$$A = \begin{bmatrix} 0 & -\tilde{C} \\ 0 & \tilde{A} \end{bmatrix}, \; B_2 = \begin{bmatrix} 0 \\ \tilde{B} \end{bmatrix}, \; B_{\mathbb{r}} = \begin{bmatrix} I_p \\ 0 \end{bmatrix}. \tag{4}$$

Note that if the matrix pair $(\tilde{A},\ \tilde{B})$ is controllable and the matrix triplet $(\tilde{A},\ \tilde{B},\ \tilde{C})$ has no zeros at the origin, it can be shown that $(A,\ B_2)$ is controllable [11].

Consider a linear switching surface as

$$\mathcal{S} = \{x :\ \sigma(t) \triangleq Sx(t) = 0\}, \tag{5}$$

where $S \in \mathbb{R}^{m\times n}$ is the full-rank-sliding matrix to be designed later so that the associated reduced-order-sliding motions have suitable dynamics.

Let us consider the following controller:

$$u(t) = -(SB_2)^{-1}(SA - \Phi S)x(t) + \vartheta(t), \tag{6}$$

where $\Phi \in \mathbb{R}^{m\times m}$ is a stable matrix and $\vartheta(t) \in \mathbb{R}^m$ denotes the nonlinear part of the controller with the following form:

$$\vartheta(t) = -(SB_2)^{-1}\rho(t)\frac{\sigma(t)}{\|\sigma(t)\|} \quad \text{if } \sigma(t) \neq 0, \tag{7}$$

in which the scalar function $\rho(\cdot)$ satisfies $\|\rho(t)\| \geq \|SB_2 f(t)\|$; for example, see [2]. It is worth noting that the term $(SB_2)^{-1}\Phi Sx(t)$ in the controller Eq. (6) is to govern the convergence rate of the system state to the sliding manifold in association with the nonlinear controller. Further, $-(SB_2)^{-1}SA$ is the so-called equivalent control necessary to maintain sliding in the absence of uncertainty. Here, similar to [3], it is assumed that $\Phi = \lambda I_m$, where $\lambda < 0$ is a given constant value. Note that unlike in [3], λ can also belong to the spectrum of A. Because we set $\Phi = \lambda I_m$, the control law $u(t)$ in Eq. (6) can be written as

$$u(t) = (SB_2)^{-1}SA_\lambda x(t) + \vartheta(t), \tag{8}$$

where $A_\lambda = \lambda I_n - A$. Now assuming that there is no matched uncertainty in Eq. (3) and letting $\rho(\cdot) \to 0$, we can consider that the controller in Eq. (8) contains only the linear part. Hence,

$$\begin{aligned} \dot{x}(t) &= Ax(t) + B_2u(t) + B_{\mathbb{r}}\mathbb{r}(t) + B_1w(t) \\ z_2(t) &= C_2x(t) + D_2u(t) \\ z_\infty(t) &= C_\infty x(t) + D_\infty u(t), \\ u(t) &= (SB_2)^{-1}SA_\lambda x(t), \end{aligned} \tag{9}$$

where $w(t)$ is a fictitious exogenous disturbance, $z_2(t) \in \mathbb{R}^{q_1}$ and $z_\infty(t) \in \mathbb{R}^{q_2}$ are the $\mathcal{H}_2$ performance output vector and $\mathcal{H}_\infty$ performance output vector of the system, respectively. The matrices in Eq. (9) are constant and of appropriate dimensions. Without the loss of generality, it is also assumed that $m \leq q_i \leq n$, $i = 1, 2$. Now, the objective can be regarded as finding a sliding matrix S so that the resulting reduced-order motion, when restricted to $\mathcal{S}$, is stable and meets $\mathcal{H}_2/\mathcal{H}_\infty$ performance specifications. Indeed, we need to choose S, with a given $\lambda < 0$, so that the obtained reduced-order-sliding mode

- guarantees $\|T_{wz_2}\|_2^2 < \delta$, where $\|T_{wz_2}\|_2$ is the $\mathcal{H}_2$ norm of the closed-loop transfer function from $w(t)$ to $z_2(t)$ and $\delta > 0$ is a predetermined closed-loop $\mathcal{H}_2$ performance, and
- minimizes the $\mathcal{H}_\infty$ performance, subject to the above item.

For this purpose, one may resort to solve a $\mathcal{H}_2/\mathcal{H}_\infty$-state-feedback problem and thereby find the switching matrix associated with the derived optimal state-feedback gain (say F). Broadly speaking, this simple scheme may not necessarily result in any solution, unless the obtained state-feedback gain F can ensure that m of the closed-loop poles are exactly located at λ. Hence, in order to design an $\mathcal{H}_2/\mathcal{H}_\infty$-based SMC, we need to address the following two problems:

1. Blend the mixed $\mathcal{H}_2/\mathcal{H}_\infty$ problem with the eigenstructure assignment method, that is, design a state-feedback F enforcing $\mathcal{H}_2/\mathcal{H}_\infty$ constraints while ensuring that m poles of the closed-loop system are precisely located at λ.
2. Obtain the sliding matrix S associated with the particular state-feedback F, derived in the previous step.

The abovementioned problems are dealt with in the following two sections.

Remark 1. *Note that the linear part of the control law can be considered as*

$$u(t) = [F_{\mathbb{r}} \, \tilde{F}] \begin{bmatrix} \xi(t) \\ \tilde{x}(t) \end{bmatrix} \triangleq Fx(t), \tag{10}$$

where $\tilde{F} \in \mathbb{R}^{m\times\tilde{n}}$ *is the state-feedback gain and* $F_{\mathbb{r}} \in \mathbb{R}^{m\times p}$ *is the feed-forward gain due to the reference signal* $\mathbb{r}(t)$.

The following lemma is recalled from [12], which will be useful in the sequel of this chapter.

Lemma 1 [12]. *The following two statements are equivalent:*

1. $\Psi + S + S^T < 0.$

2. *The following LMI is feasible with respect to* U.

$$\begin{bmatrix} \Psi + P - (U + U^T) & S^T + U^T \\ S + U & -P \end{bmatrix} < 0,$$

where P *is a positive definite matrix.*

It should be noted that Lemma 1 provides a necessary and sufficient condition. However, while imposing some constraints (e.g., structural constraints) on U, the sufficiency of the lemma is not violated; that is, always (2) ⇒ (1).

3. Partial eigenstructure assignment for the design of $\mathcal{H}_2/\mathcal{H}_\infty$-based SMC

3.1. LMI characterizations

We need to consider the state-feedback synthesis with a combination of $\mathcal{H}_2/\mathcal{H}_\infty$ performance specifications. In what follows, to avoid the conservatism introduced by the so-called *quadratic* approach for the design of feedback gains with respect to $\mathcal{H}_2/\mathcal{H}_\infty$ performance specifications, we need to recall the so-called *extended* LMI methods developed for the $\mathcal{H}_2$ and $\mathcal{H}_\infty$ control problems from, for example, [12, 13]. This form of LMI characterization will also be shown to be very useful for the novel SMC of this chapter, as it provides us with the possibility to design a certain partial eigenstructure assignment scheme which can ensure precise locations for some of the closed-loop system poles.

3.1.1. $\mathcal{H}_2$ LMI characterization

The $\mathcal{H}_2$ control synthesis problem, by assuming the control law as $u(t) = Fx(t)$, can be addressed through the following optimization problem [12]:

$$\text{Minimize } \delta \text{ subject to} \tag{MH2}$$

$$\begin{bmatrix} -(G+G^T) & \star & \star & \star \\ AG + B_2Y + X_i & -X_i & \star & \star \\ C_2G + D_2Y & 0 & -\delta I & \star \\ G & 0 & 0 & -X_i \end{bmatrix} < 0, \tag{11}$$

$$\begin{bmatrix} -Z & \star \\ B_1 & -X_i \end{bmatrix} < 0, \tag{12}$$

$$\text{trace}(Z) < 1, \tag{13}$$

with respect to decision variables X_i, $i = 1, \cdots, \mathcal{N}$, Z, Y, and G, where X_i and Z are s.p.d matrices. $\mathcal{N}$ hereafter denotes the number of constraints and thus the independent Lyapunov variables. As $G + G^T > 0$, G will be invertible and the state feedback is obtained as $F = YG^{-1}$.

3.1.2. $\mathcal{H}_\infty$ LMI characterization

Given scalar $0 < \nu \ll 1$, the $\mathcal{H}_\infty$ problem, by assuming the control law as $u(t) = Fx(t)$, can be set as the following minimization problem [13].

$$\text{Minimize } \gamma \text{ subject to} \tag{MHI}$$

$$\begin{bmatrix} X_i - (G+G^T) & \star & \star & \star \\ G + \nu(AG + B_2Y) & -X_i & \star & \star \\ C_\infty G + D_\infty Y & 0 & -\nu^{-1}I & \star \\ 0 & B_1 & 0 & -\gamma\nu^{-1}I \end{bmatrix} < 0, \tag{14}$$

with respect to decision variables $X_i > 0$, $i = 1, \cdots, \mathcal{N}$, Y, G, and $\gamma > 0$. Again, the state feedback is obtained as $F = YG^{-1}$.

Remark 2. *It is worth mentioning that the advantage of both LMIs* (11) *and* (14) *lies in the fact that the product terms between the matrix A and the Lyapunov matrices X_i disappear which is particularly useful for a wide range of applications such as mixed $\mathcal{H}_2/\mathcal{H}_\infty$-feedback gain design and cases where the system matrices belong to a given polytopic region. Besides, as seen from* Eqs. (11) *and* (14), *the controller is not dependant on the Lyapunov matrix, but rather the new introduced matrix variable G.*

3.1.3. Mixed $\mathcal{H}_2/\mathcal{H}_\infty$ state feedback using improved LMI characterizations

An interesting application of the mentioned so-called *extended* LMI characterizations for $\mathcal{H}_2$ and $\mathcal{H}_\infty$ is the mixed $\mathcal{H}_2/\mathcal{H}_\infty$ state-feedback problem. The aim is to design-feedback gains such that they

- ensure the $\mathcal{H}_2$ performance which means that for a prescribed closed-loop $\mathcal{H}_2$ performance $\delta > 0$, we have $\| T_{wz_2} \|_2^2 < \delta$;
- minimize the $\mathcal{H}_\infty$ performance, subject to the above constraint.

This problem can be formulated through an LMI program in decision variables $X_i > 0$, $i = 1, \cdots, \mathcal{N}$, $Z > 0$, Y, G, and $\gamma > 0$:

$$\begin{array}{c}\text{minimize} \quad \gamma \\ \text{subject to Eqs. (11), (12), (13), and (14),}\end{array} \tag{MHH}$$

where $\delta > 0$ and $0 < \nu \ll 1$ are the given scalars. Notice that another alternative for addressing the mixed $\mathcal{H}_2/\mathcal{H}_\infty$ state-feedback problem is the so-called *quadratic approach* (see, e.g., [14]), which is a well-known scheme to address the nonlinearity involved in the matrix inequalities by using a common Lyapunov matrix for all the involved objectives. However, this scheme introduces a significant conservatism to the problem in most of the practical cases. The other alternatives, such as (MHH), which are more computationally expensive, have been basically considered in the literature in order to reduce the conservatism of the quadratic approach.

Remark 3. *Another alternative for the mixed control problem is to design a feedback gain that minimizes the $\mathcal{H}_2$ norm of one channel while satisfying an $\mathcal{H}_\infty$-norm constraint on the same or another channel; see, for example, [15]. Hence, in this case, the mixed $\mathcal{H}_2/\mathcal{H}_\infty$ problem, given $\gamma > 0$ and $0 < \nu \ll 1$, can be set as follows:*

$$\begin{array}{c}\textit{minimize} \quad \delta \\ \textit{subject to } (11), (12), (13) \textit{ and } (14),\end{array} \tag{MHHN}$$

where $X_i > 0$, $i = 1, \cdots, \mathcal{N}$, $Z > 0$, Y, G, and $\delta > 0$ are decision variables.

3.2. Partial eigenstructure assignment problem

Locating exactly m poles at a specific location can fortunately be performed through the LMI characterization presented in the previous section. Our specific partial assignment of the set of eigenvalues

$$\{\overbrace{\lambda, \cdots, \lambda}^{m \text{ times}}\}, \tag{15}$$

by state feedback can be implemented in two steps:

1. compute the base $\begin{bmatrix} M_\lambda \\ N_\lambda \end{bmatrix}$ of null space of $[A - \lambda I \; B_2]$ with conformable partitioning;
2. with arbitrary $\eta_i \in \mathbb{R}^m$, $i = 1, \cdots, m$, the state feedback will be obtained as $F = YG^{-1}$ with

$$Y = N\Sigma_N, G = M\Sigma_M, \tag{16}$$

in which

$$\begin{aligned} N &:= [\overbrace{N_\lambda, \cdots, N_\lambda}^{m \text{ times}}, \overbrace{I, \cdots, I}^{(n-m) \text{ times}}], \\ M &:= [\overbrace{M_\lambda, \cdots, M_\lambda}^{m \text{ times}}, \overbrace{I, \cdots, I}^{(n-m) \text{ times}}], \\ \Sigma_N &:= \text{diag}(\eta_1, \cdots, \eta_m, \kappa_1, \cdots, \kappa_{(n-m)}), \\ \Sigma_M &:= \text{diag}(\eta_1, \cdots, \eta_m, \iota_1, \cdots, \iota_{(n-m)}) \end{aligned} \tag{17}$$

with $\kappa_i \in \mathbb{R}^n$ and $\iota_i \in \mathbb{R}^n$. Note that only vectors η_k are related to the assignment of the m eigenvalues to λ. In other words, other vectors (κ_k and ι_k) are not exploited in the pole-placement purposes and thereby can be employed to meet other Lyapunov-type constraints.

Now, provided by the LMI characterization, for example, (MHH), the first step of our $\mathcal{H}_2/\mathcal{H}_\infty$-based SMC design can be set as an LMI program in the variables $X_i > 0$, $i = 1, \cdots, \mathcal{N}$, Σ_M, Σ_N, and $\gamma > 0$, by recasting (MHH) as:

$$\begin{aligned} &\text{minimize} \ \ \gamma \\ &\text{subject to (11), (12), (13), (14), and (16).} \end{aligned} \tag{MHH2}$$

However, we have not yet shown that the set of closed-loop eigenvalues encompasses Eq. (15). This is the subject of the following lemma.

Lemma 2. *Set* (15) *is a subset of the closed-loop system eigenvalues, acquired by applying the state feedback $F = YG^{-1}$ with Y and G presented in Eq.* (16), *to the system in Eq.* (3) *in the absence of uncertainty, that is, $f = 0$.*

Proof. Using Eq. (16), we can write

$$\begin{aligned} &(A + B_2F)M_\lambda\eta_i \\ &= \left[A + B_2(N\Sigma_N)(M\Sigma_M)^{-1}\right]M_\lambda\eta_i \\ &= \left[A + B_2(N\Sigma_N)(M\Sigma_M)^{-1}\right](M\Sigma_M)e_i \\ &= [A(M\Sigma_M) + B_2(N\Sigma_N)]e_i \\ &= AM_\lambda\eta_i + B_2N_\lambda\eta_i \\ &= \lambda M_\lambda\eta_i \qquad\qquad i = 1, \cdots, m. \end{aligned}$$

Note that e_k here denotes the canonical basis of $\mathbb{R}^n$. □

4. Deriving the switching-function matrix

This subsection proposes two approaches to obtain the sliding matrix S associated with the state feedback F, derived in the previous subsection based on the partial eigenstructure assignment scheme.

4.1. Approach 1

The first approach is represented in the following theorem.

Theorem 1. *Let (A, B_2) be a controllable matrix pair. Then*

i. $\forall\, \lambda \in \mathbb{R}_-$, *there always exists a gain matrix F so that m of the eigenvalues of $A + B_2F$ are equivalent to λ, and $A + B_2F$ has m-independent eigenvectors associated with λ.*

ii. *Define $S = [v_1,\cdots,v_m]^T$, where v_i is a left eigenvector of $A + B_2F$ associated with the eigenvalue λ, then, $S(A_\lambda - B_2F) = 0$ and SB_2 is invertible.*

Proof. (i) As (A,B_2) is controllable, we can claim that $(\lambda I - A,B_2)$ is also controllable for any $\lambda \in \mathbb{R}_-$. Then, it is easy to see that we can always find F such that the null space of $A_\lambda - B_2F$ has dimension m, which implies that $A + B_2F$ has m-independent eigenvectors associated with λ.

(ii) Define $S = [v_1,\cdots,v_m]^T$, it is easy to show that $S(A_\lambda - B_2F) = 0$. Now, assume

$$SB_2 := \begin{bmatrix} v_1^T \\ \vdots \\ v_m^T \end{bmatrix} B_2 = \Omega,$$

where $\Omega \in \mathbb{R}^{m\times m}$. If Ω is not full rank, then there exists a nonsingular matrix Λ such that the first row of $\Lambda\Omega$ is zero. This is equivalent to

$$\Lambda \begin{bmatrix} v_1^T \\ \vdots \\ v_m^T \end{bmatrix} B_2 := \begin{bmatrix} \tilde{v}_1^T \\ \vdots \\ \tilde{v}_m^T \end{bmatrix} B_2 = \Lambda\Omega,$$

that is, there exists a vector $\tilde{v}_1$ such that $\tilde{v}_1^T B_2 = 0$. On the other hand, we know $\tilde{v}_1^T [A_\lambda - B_2F] = 0$, and so

$$\text{rank}\Big([\lambda I - (A + B_2F)\ \ B_2]\Big) < n.$$

This is clearly in contradiction with the controllability of (A, B_2). In other words, if we can find a left eigenvector of $A + B_2F$ associated with λ that is orthogonal to B_2, (A, B_2) must be uncontrollable, which is obviously a contradiction.

In brief, by virtue of Theorem 1, the switching-function matrix S, associated with the state feedback F, obtained through solving the LMI problem in (MHH2), can be selected as the set of m linearly independent left eigenvectors of $A + B_2F$ associated with the (arbitrarily selected) repeated eigenvalue $\lambda \in \mathbb{R}_-$.

4.2. Approach 2

An alternative approach to obtain the sliding matrix is to address the equality

$$(SB_2)^{-1} SA_\lambda = F, \tag{18}$$

utilizing an LMI optimization method as follows. As the matrix S must ensure the invertibility of SB_2, let us suppose $S = B_2^T P$, with P an s.p.d matrix which will be obtained hereafter. The condition in Eq. (18) can be dealt with a simple relaxation method as

$$\text{minimize}\ \alpha\ \ \text{subject to}\ \| B_2^T P(A_\lambda - B_2 F) \| < \alpha,$$

where $\alpha > 0$ is a scalar variable and F is a given state-feedback matrix, obtained in the previous subsection, ensuring m of the closed-loop eigenvalues are equal to λ. Simply it can be shown that the above problem is equivalent to the following LMI minimization problem:

$$\text{minimize}\ \alpha\ \ \text{subject to} \begin{bmatrix} -\alpha I & \star \\ B_2^T P(A_\lambda - B_2 F) & -\alpha I \end{bmatrix} < 0. \tag{19}$$

Hence, the $\mathcal{H}_2/\mathcal{H}_\infty$-based SMC problem is now to find the global solution of the above minimization problem, and then the sliding matrix is obtained as $S = B_2^T P$. In the case of feasibility, this problem will enforce α so that it is an extremely small number associated with the precision of the computational unit.

Notice that this approach is numerically very efficient and attractive compared to the first approach.

5. The summary of $\mathcal{H}_2/\mathcal{H}_\infty$-based SMC design method

Now, we summarize the proposed $\mathcal{H}_2/\mathcal{H}_\infty$-based SMC in the following theorem.

Theorem 2. *Assume that the optimization problem in* (MHH2) *has a solution F for some $\delta > 0$ and $\gamma > 0$. Then, the $\mathcal{H}_2/\mathcal{H}_\infty$ performance constraints $\| T_{wz_2} \|_2^2 < \delta$ and $\| T_{wz_\infty} \|_\infty^2 < \gamma$ are ensured, and after the reaching time t_s, the resulting reduced $n - m$-order-sliding mode dynamics, obtained by applying the following control law:*

$$u(t) = Fx(t) + \vartheta(t), \tag{20}$$

where $\vartheta(t)$ is introduced in Eq. (7), *to system* (3), *is asymptotically stable.*

Proof. Consider a change of coordinates $x \mapsto T_r x$. In this new coordinate system, the new matrix pair $(\overline{A}, \overline{B}_2)$ is of the form

$$\overline{A} = \begin{bmatrix} A_{11} & A_{12} \\ A_{21} & A_{22} \end{bmatrix}, \overline{B}_2 = \begin{bmatrix} 0 \\ B_p \end{bmatrix} \tag{21}$$

where the square matrix $B_p \in \mathbb{R}^{m \times m}$ has full rank and more importantly is nonsingular; see [1]. Suppose also that $\overline{F}$ is the state feedback in the new coordinate that ensures the closed-loop stability, assigns m poles of the closed-loop system at λ, and satisfies the $\mathcal{H}_2/\mathcal{H}_\infty$ performance constraints $\| T_{wz_2} \|_2^2 < \delta$ and $\| T_{wz_\infty} \|_\infty^2 < \gamma$. Now, let the switching-function matrix in the original coordinates be parameterized such that [1]

$$S = S_2[-\mathcal{M} \quad I_m]T_r, \tag{22}$$

where $S_2 \in \mathbb{R}^{m\times m}$. Notice that theoretically the choice of S_2 may not influence the sliding motion [1]. According to the discussion given in the previous section, it can be readily shown that there exists a matrix $\mathcal{M}$ such that $\overline{F} = (\overline{S}\overline{B}_2)^{-1}\overline{S}(\overline{A} - \lambda I_n)$, where $\overline{S} = S_2[-\mathcal{M} \quad I_m]$ denotes the switching-function matrix in the new coordinate. Let $(\overline{x}_1, \overline{x}_2)$ be the partition of the system states associated with the certain system coordinates in Eq. (21), then it can be shown that while the system states are confined to the sliding manifold, that is, $\sigma = 0$, the reduced-order-sliding mode dynamics are governed by the stable reduced-order system matrix $A_{11} + A_{12}\mathcal{M}$.

Moreover, the dynamics of σ can be derived by taking the time derivative of Eq. (5), substituting in the state equation (3), and using controllers (8) and (7), that is,

$$\dot{\sigma}(t) = \lambda\sigma(t) - \rho(t)\frac{\sigma(t)}{\|\sigma(t)\|} + SB_2 f(t). \tag{23}$$

Finally, it follows from $\|SB_2 f(t)\| \leq \rho(t)$ that the reachability condition $\frac{\sigma^T\dot{\sigma}}{\|\sigma\|} < 0$ holds. □

6. Design of SMC with additional regional pole-placement constraints

Note that the proposed method here offers the advantage of introducing additional convex constraints on the closed-loop dynamics. By locating the closed-loop system poles in a preselected region, an adequate transient response for system trajectories can be guaranteed [14]. Therefore, the objective is to augment the optimization problems previously described by pole-clustering constraints. Note also that as it is already ensured that m of the closed-loop eigenvalues are exactly assigned to a given negative real value (λ), the remaining eigenvalues in fact belong to the spectrum of the reduced $n - m$-order-sliding motion. As a result, a satisfactory transient response for the sliding motion can be achieved by clustering the poles governing the sliding motion.

Let us have a brief introduction to the LMI region. Simply, an LMI region is a subset $\mathcal{D}$ of the complex plane as

$$\mathcal{D} := \{z \in \mathrm{r\}E502} : f_{\mathcal{D}}(z) \triangleq \Xi + z\Pi + \bar{z}\Pi^T < 0\} \tag{24}$$

in which $\Xi = \Xi^T \in \mathbb{R}^{\xi\times\xi}$ and $\Pi \in \mathbb{R}^{\xi\times\xi}$ are real matrices. $f_{\mathcal{D}}(z)$ is also called the characteristic equation of the region $\mathcal{D}$.

Definition 1 [16]. *A real matrix $\mathcal{A}$ is said to be $\mathcal{D}$-stable if all its eigenvalues lie within the LMI region $\mathcal{D}$.*

Lemma 3 [16]. *A real matrix $\mathcal{A}$ is said to be $\mathcal{D}$-stable if a symmetric matrix $X_{\mathcal{D}} > 0$ exists so that*

$$\Xi \otimes X_{\mathcal{D}} + \Pi \otimes (X_{\mathcal{D}}\mathcal{A}) + \Pi^T \otimes (\mathcal{A}^T X_{\mathcal{D}}) < 0, \tag{25}$$

where $\otimes$ denotes the Kronecker product.

However, the synthesis problem obtained by imposing the pole-clustering constraints presented in, for example, [14] or [16] to the synthesis problem in (MHH2) would not result in a convex problem. Alternatively, the regional pole-clustering constraints can be reformulated so that the product term between the Lyapunov matrix X_i and the system matrix A is removed.

An instrumental theorem is represented first and the main theorem will be presented later in Theorem 4.

Theorem 3. *Let $\mathcal{A}$ be a real matrix. The following statements are equivalent, with s.p.d X, $\mathcal{G}$ and given real matrices $0 < \Xi \in \mathbb{R}^{\xi\times\xi}$ and $\Pi \in \mathbb{R}^{\xi\times\xi}$.*

1. *$\mathcal{A}$ is $\mathcal{D}$-stable, where $\mathcal{D}$ is given in Eq. (24).*

2. *$\exists$ X such that*

$$\begin{bmatrix} \Xi\otimes X + \Pi\otimes(X\mathcal{A}) + \Pi^T\otimes(X\mathcal{A})^T & \star \\ 0 & -X \end{bmatrix} < 0.$$

3. *$\exists$ $X > 0$ and $\mathcal{G}$ such that*

$$\begin{bmatrix} -(\mathcal{G}+\mathcal{G}^T) & \star & \star & \star \\ (\Pi\otimes\mathcal{A})\mathcal{G} + I_\xi\otimes X & -I_\xi\otimes X & \star & \star \\ \mathcal{G} & 0 & -I_\xi\otimes X & \star \\ \mathcal{G} & 0 & 0 & -\Xi^{-1}\otimes X \end{bmatrix} < 0. \tag{26}$$

Proof. Refer to the Appendix. □

While Eq. (26) can be seen as a necessary and sufficient condition for $\mathcal{D}$-stability, it is not very useful in terms of control synthesis purposes. Further, since $\Xi = 0$, the result of Theorem 3 cannot cover the standard continuous-time systems stability. However, if we let $\mathcal{G} = I_\xi \otimes G$ in Eq. (26), a sufficient condition is achieved which is beneficial for the control synthesis purposes.

Theorem 4. *Let $\mathcal{A}$, $0 \le \Xi \in \mathbb{R}^{\xi\times\xi}$, and $\Pi \in \mathbb{R}^{\xi\times\xi}$ be real matrices. $\mathcal{A}$ is $\mathcal{D}$-stable if*

$$\begin{bmatrix} -I_\xi\otimes(G+G^T) & \star & \star & \star \\ \Pi\otimes(\mathcal{A}G) + I_\xi\otimes X & -I_\xi\otimes X & \star & \star \\ I_\xi\otimes G & 0 & -I_\xi\otimes X & \star \\ \Xi^{\frac{1}{2}}\otimes G & 0 & 0 & -I_\xi\otimes X \end{bmatrix} < 0, \tag{27}$$

where G is a general matrix and X is an s.p.d matrix.

Proof. The proof can be performed similar to the proof of Theorem 3 by letting $\mathcal{G} = I_\xi \otimes G$. □

Clearly, the above theorem does not require $\Xi > 0$, but $\Xi \ge 0$. This is indeed a generalization of the extended Lyapunov theorem presented in Theorem 3.1 of [12], and the usual stability region can be obtained by letting $\Xi = 0$ and $\Pi = 1$ in Eq. (27):

$$\begin{bmatrix} -(G+G^T) & \star & \star \\ \mathcal{A}G+X & -X & \star \\ G & 0 & -X \end{bmatrix} < 0. \quad (28)$$

Moreover, the equivalence of Eq. (28) to the standard Lyapunov stability inequality for continuous-time linear systems is presented in [12].

Specifically, let us confine the closed-loop poles to the region r}E512(α, r, θ) (see [14]) which can ensure a minimum decay rate α, a minimum damping ratio $\zeta = \cos\theta$, and a maximum undamped natural frequency $\omega_d = r\sin\theta$. The LMI region for an αstability, that is, $\mathrm{Re}(z) < -\alpha$, can be obtained through Eq. (27), with $\Xi = 2\alpha$, $\Pi = 1$, $\mathcal{A}A + B_2F$, and XX_i. Moreover, by letting $\Xi = 0$ and $\Pi = \begin{bmatrix} \sin\theta & \cos\theta \\ -\cos\theta & \sin\theta \end{bmatrix}$, the LMI region for a conic sector r}E512$(0,0,\theta)$ is achieved. Eventually, a disk centered at the origin with radius r corresponds to

$$\Xi = \begin{bmatrix} -r & 0 \\ 0 & -r \end{bmatrix}, \Pi = \begin{bmatrix} 0 & 1 \\ 0 & 0 \end{bmatrix}. \quad (29)$$

However, for this special pole-clustering constraint, as Ξ is not a semi-positive definite matrix, the LMI region cannot be obtained through Eq. (27). We can alternatively state the following theorem.

Theorem 5. *Let $\mathcal{A}$ be a real matrix. The following conditions are equivalent:*

1. *The eigenvalues of $\mathcal{A}$ lie in a disk centered at the origin with radius r.*
2. *There exists a symmetric matrix $X > 0$ such that*

$$\frac{1}{r}\mathcal{A}X\mathcal{A}^T - rX < 0. \quad (30)$$

3. *There exists a symmetric matrix $X > 0$ such that*

$$\begin{bmatrix} -rX & \star \\ X\mathcal{A}^T & -rX \end{bmatrix} < 0. \quad (31)$$

4. *There exist a symmetric matrix $X > 0$, and a matrix G such that*

$$\begin{bmatrix} -rX & \star \\ G^T\mathcal{A}^T & -(G+G^T) + \frac{1}{r}X \end{bmatrix} < 0. \quad (32)$$

Proof. Refer to the Appendix. □

Notice that the above theorem with $r = 1$ reduces to the standard Lyapunov stability inequality for discrete-time linear systems and its extended (robust) version; for example, see [17]. Now, the extended LMI region for a disk centered at the origin with radius r is as follows:

$$\begin{bmatrix} -rX_i & \star \\ (AG + B_2Y)^T & -(G+G^T) + \frac{1}{r}X_i \end{bmatrix} < 0, \quad (33)$$

which is obtained by replacing $\mathcal{A}{:=}A + B_2F$, $X{:=}X_i$ in Eq. (32) and introducing $Y = FG$.

Remark 4. *Exploiting a common G may also lead to conservatism compared with the methods, for example, in [18]. However, the methods in the aforementioned references are not beneficial for the control synthesis aims, unless gain-scheduled controllers [19] are considered. Moreover, by employing two instrumental variables, a different sufficient condition for robust $\mathcal{D}$ stability has been developed in [20] which is not applicable to the continuous-time control synthesis purposes. Nevertheless, the approach here can achieve less conservative results through employing non-common Lyapunov variables for every involved specification.*

7. Numerical examples

This section evaluates the effectiveness of the proposed theory using a numerical example. Consider a two-input, two-output, fourth-order plant describing the motion of a Boeing B-747 aircraft obtained by linearization around an operating condition of 20,000 ft. altitude with a speed of Mach 0.8 [21]. The system matrices are as follows:

$$\tilde{A} = \begin{bmatrix} -0.1196 & 0.0004 & -1.0001 & 0.0383 \\ -4.1195 & -0.9743 & 0.2919 & -0.0004 \\ 1.6204 & -0.0161 & -0.2320 & -0.0001 \\ 0.0007 & 1.0054 & 0.0003 & 0.0003 \end{bmatrix},$$

$$\tilde{B} = \begin{bmatrix} -0.0004 & 0.0126 \\ 0.3103 & 0.1832 \\ 0.0124 & -0.9219 \\ -0.0001 & -0.0002 \end{bmatrix}, \tilde{C} = \begin{bmatrix} 1 & 0 & 0 & 0 \\ 0 & 0 & 0 & 1 \end{bmatrix},$$

and the system state, output, and input vectors are

$$\begin{aligned} \tilde{x}(t) &= [\beta(t) \quad p(t) \quad r(t) \quad \phi(t)]^T, \\ \tilde{y}(t) &= [\beta(t) \quad \phi(t)]^T, \\ u(t) &= [\delta_a(t) \quad \delta_r(t)]^T. \end{aligned}$$

where $\beta(t)$, $p(t)$, $r(t)$, $\phi(t)$, $\delta_a(t)$, and $\delta_r(t)$ denote the sideslip angle, the roll rate, the yaw rate, the roll angle, the aileron deflection, and the rudder deflection, respectively.

We also let

$$\begin{aligned} C_2 &= \begin{bmatrix} \text{diag}(0.1, 0.1, 10, 10, 1, 1) \\ 0_{2\times 6} \end{bmatrix}, \\ D_2 &= \begin{bmatrix} 0_{6\times 2} \\ \text{diag}(1, 1) \end{bmatrix}, \\ B_1 &= I_6. \end{aligned}$$

Note that the last two nonzero terms of C_2 are associated with the integral action and are less heavily weighted. In addition, the third and fourth terms of C_2 have strongly been weighted in comparison with the fifth and sixth terms to provide an adequate quick closed-loop response in terms of the angular acceleration in roll and yaw. We also aim to assign the closed-loop poles in the half-plane $x < -\alpha < -0.1$.

We solve the minimization problem in (MH2), with $\lambda = -3$, and the state-feedback gain is obtained as

$$F = \begin{bmatrix} 0.7166 & 68.7237 & 16.0446 & -19.6616 & -4.0591 & -62.2050 \\ 34.5978 & 24.6097 & -23.0976 & 0.7777 & 7.4387 & -5.7251 \end{bmatrix}. \tag{34}$$

Employing the first proposed approach in Section 4, the associated sliding function matrix for the augmented system is

$$S = \begin{bmatrix} -0.7351 & -0.6091 & 0.2907 & 0.0151 & -0.0621 & -0.0066 \\ -0.4278 & 0.6381 & 0.2355 & -0.1309 & -0.0638 & -0.5772 \end{bmatrix}. \tag{35}$$

The sliding motion is governed by the set of poles $\{-2.3205 \pm 3.0365i, -1.9203 \pm 1.2377i\}$, and the associated true value of $\mathcal{H}_2$ cost from w to z is 28.0959. Assuming the matched

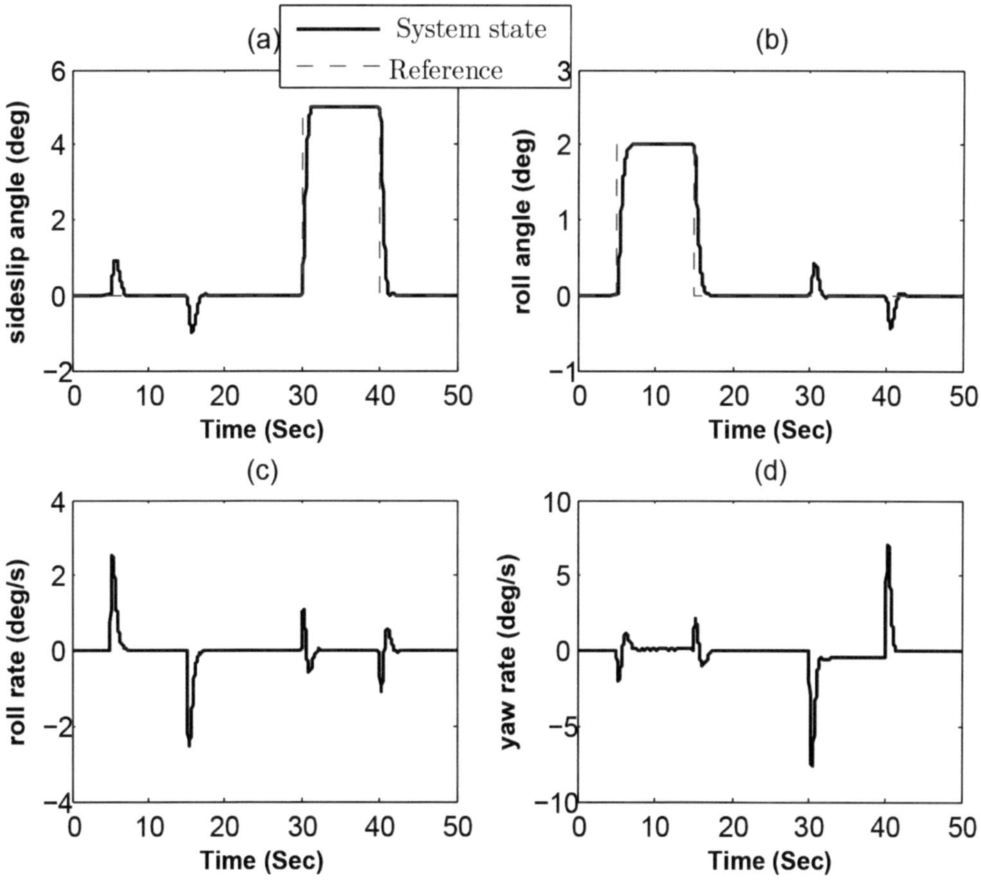

Figure 1. System states.

uncertainty term in Eq. (1) as $f(t) = \begin{bmatrix} 0.2\sin(t)\beta(t) \\ 0.3\sin(t)\phi(t) \end{bmatrix}$, using the proposed SMC with the obtained linear gain F in Eq. (34) and the associated switching-function matrix F in Eq. (35), and letting the switching gain $\rho = 1$, and considering a step of 5° for β during 30–40 s as well as a step of 2° for φ during 5–15 s, **Figures 1–3** show the tracking responses of the system. Note that the discontinuity in the nonlinear control term $\vartheta(t)$ in Eq. (7) is smoothed by using a sigmoidal approximation [11] as

$$\vartheta_\varepsilon(t) = -(SB_2)^{-1}\rho(t)\frac{\sigma(t)}{\varepsilon + \|\sigma(t)\|} \tag{36}$$

with the scalar $\varepsilon = 0.01$ and $\rho(t) = 1$, where this can remove the discontinuity at $\sigma = 0$ and introduce the possibility to accommodate the actuator rate limits.

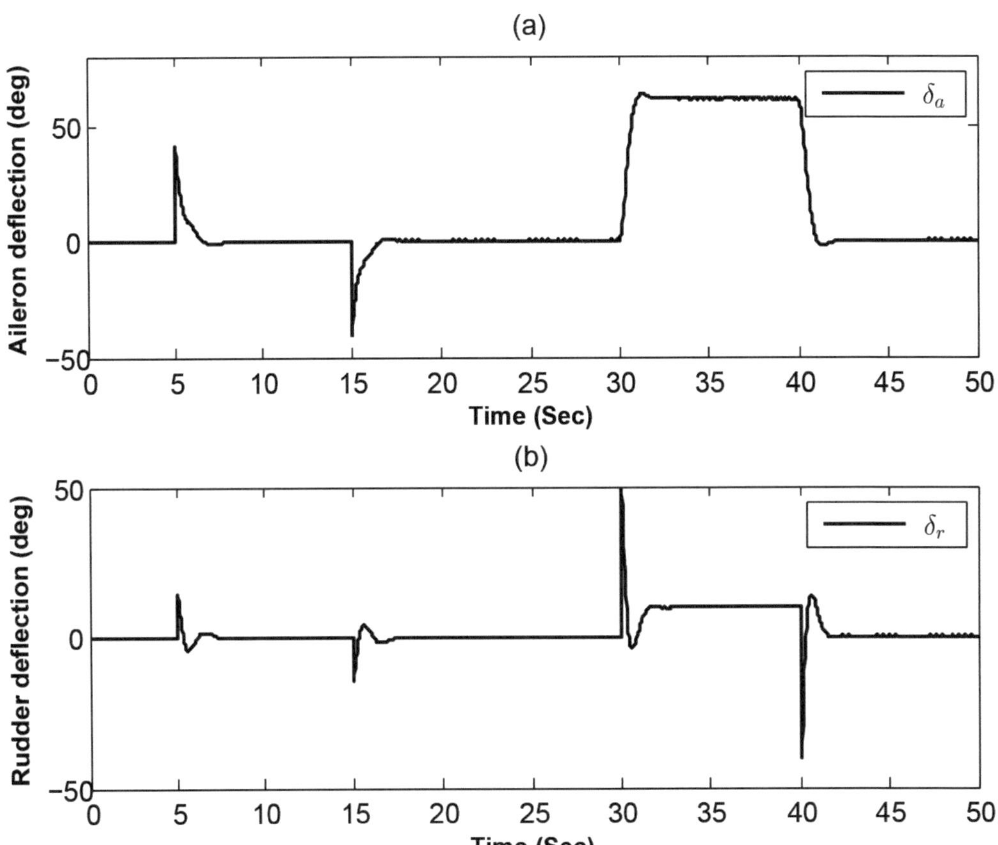

Figure 2. Control efforts.

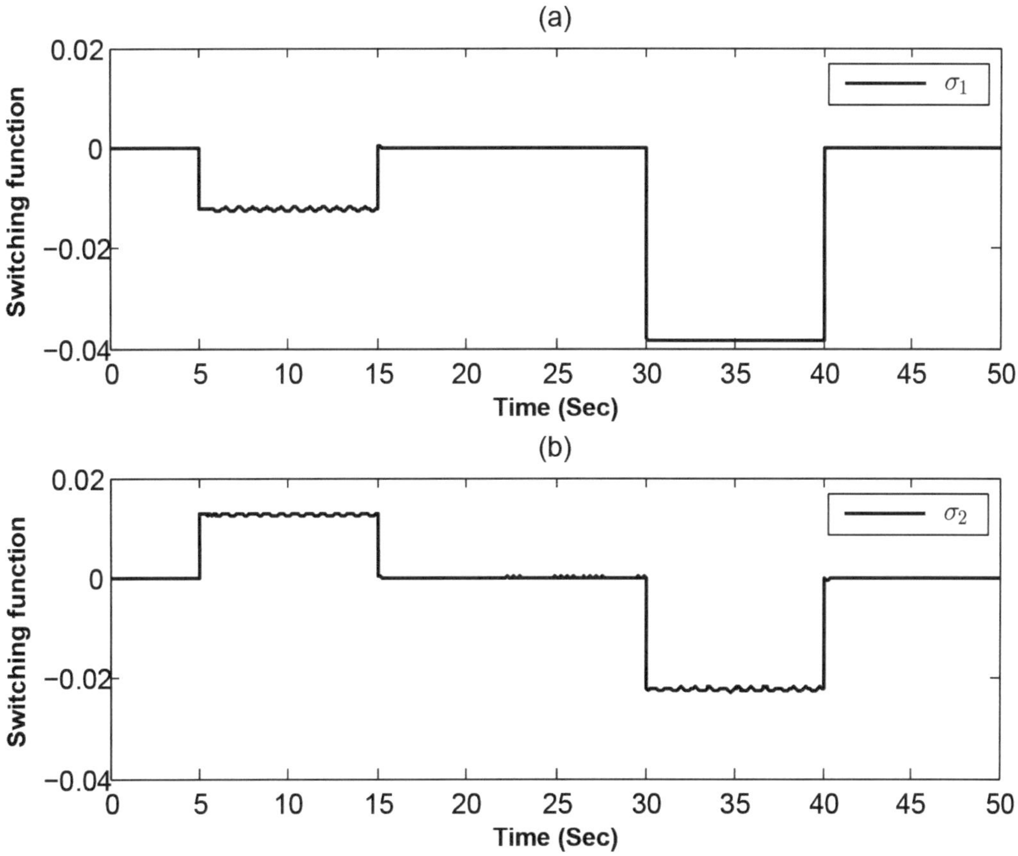

Figure 3. Switching function.

8. Conclusions

The focus of this chapter was on the development of novel framework for designing a sliding surface for a given system while enforcing a number of Lyapunov-type constraints such as the $\mathcal{H}_2/\mathcal{H}_\infty$ and/or regional pole clustering. We specifically considered the problem of output tracking using a suboptimal state-feedback SMC. In doing so, in the first stage, through a convex optimization approach, a state-feedback gain is designed while assigning a certain number (m) of the closed-loop system eigenvalues to a predetermined value, as well as satisfying $\mathcal{H}_2/\mathcal{H}_\infty$-norm constraints. The advantages of the proposed scheme are threefold: (i) it can set the stage for designing SMC while the level of control efforts is taken into account; (ii) it makes it possible to integrate a number of Lyapunov-type constraints, for example, regional pole-placement constraints, into the SMC design problem; and (iii) the controller can be computed in a numerically very efficient method. The achieved results confirmed the effectiveness of the proposed approach.

A. Proof of Theorem 3

Notice that the equivalence between Eqs. (1) and (2) can be obtained from Lemma 3. We will show the equivalence between Eqs. (2) and (3) here. The use of Lemma 1 with $\Psi = \Xi \otimes \overline{X}$, $U = \overline{\mathcal{G}}$ and $S = \Pi \otimes (\overline{X}\mathcal{A})$, with $\overline{X} = X^{-1}$, yields

$$\begin{bmatrix} P - (\overline{\mathcal{G}} + \overline{\mathcal{G}}^T) + \Xi \otimes \overline{X} & \star \\ \Pi \otimes (\overline{X}\mathcal{A}) + \overline{\mathcal{G}} & -P \end{bmatrix} < 0,$$

or equivalently,

$$\begin{bmatrix} P - (\overline{\mathcal{G}} + \overline{\mathcal{G}}^T) + \Xi \otimes \overline{X} & \star \\ \Pi \otimes \mathcal{A} + (I_\xi \otimes X)\overline{\mathcal{G}} & -(I_\xi \otimes X)P(I_\xi \otimes X) \end{bmatrix} < 0, \tag{37}$$

By performing the congruence transformation $\begin{bmatrix} \mathcal{G} & 0 \\ 0 & I \end{bmatrix}$, with $\mathcal{G} = \overline{\mathcal{G}}^{-1}$, and using the Schur complement, Eq. (37) becomes

$$\begin{bmatrix} -(\mathcal{G} + \mathcal{G}^T) & \star & \star & \star \\ (\Pi \otimes \mathcal{A})\mathcal{G} + I_\xi \otimes X & -(I_\xi \otimes X)P(I_\xi \otimes X) & \star & \star \\ \mathcal{G} & 0 & -P^{-1} & \star \\ \mathcal{G} & 0 & 0 & -\Xi^{-1} \otimes X \end{bmatrix} < 0.$$

The above inequality finally linearizes to Eq. (26) with the choice $P = I_\xi \otimes X^{-1}$.

B. Proof of Theorem 5

The equivalence between Eqs. (1) and (3) is shown in, for example, [14]. Moreover, the equivalence between Eqs. (2) and (3) is simply obtained through applying the Schur complement with respect to the block $(2, 2)$ in Eq. (31). The proof can be followed by noticing that if one applies the Schur complement with respect to the block $(1, 1)$ in Eq. (32), Eq. (30) is recovered by choosing $G = G^T = \frac{1}{r}X > 0$, hence Eq. (2) implies Eq. (4). Also, by left and right multiplying Eq. (32) by $[I \quad \mathcal{A}]$ and $[I \quad \mathcal{A}]^T$, respectively, one can achieve Eq. (30). Hence, Eq. (4) implies Eq. (2), and the proof is completed.

Author details

Ahmadreza Argha* and Steven W. Su

*Address all correspondence to: rezaargha@yahoo.com

Faculty of Engineering and Information Technology, University of Technology, Sydney, Broadway, NSW, Australia

References

[1] V. I. Utkin. Sliding modes in control optimization, communications and control engineering series. Springer-Verlag, London, 1992.

[2] C. Edwards and S. K. Spurgeon. Sliding mode control: Theory and applications. Taylor and Francis, London, 1998.

[3] C. Edwards. A practical method for the design of sliding mode controllers using linear matrix inequalities. Automatica, 40:1761–1769, 2004.

[4] G. Herrmann, S. K. Spurgeon, and C. Edwards. A robust sliding-mode output tracking control for a class of relative degree zero and non-minimum phase plants: A chemical process. International Journal of Control, 72:1194–1209, 2001.

[5] H. Choi. A new method for variable structure control system design: A linear matrix inequality approach. Automatica, 33:2089–2092, 1997.

[6] X. Han, E. Fridman, and S. K. Spurgeon. Sliding-mode control of uncertain systems in the presence of unmatched disturbances with applications. International Journal of Control, 83(12):2413–2426, 2010.

[7] H. H. Choi. Variable structure output feedback control design for a class of uncertain dynamic systems. Automatica, 38(2):335–341, 2002.

[8] P. G. Park, D. J. Choi, and S. G. Kong. Output feedback variable structure control for linear systems with uncertainties and disturbances. Automatica, 43(1):72–79, 2007.

[9] Y. Pan and K. Furuta. VSS controller design for discrete-time systems. Control-Theory and Advanced Technology, 10(4):669–687, 1994.

[10] C. Y. Tang and E. A. Misawab. Sliding surface design for discrete VSS using LQR technique with a preset real eigenvalue. Systems and Control Letters, 45(1):1–7, 2002.

[11] H. Alwi and C. Edwards. Fault tolerant control using sliding modes with on-line control allocation. In Fault tolerant flight control, pp. 247–272. Springer, 2010.

[12] P. Apkarian, H. D. Tuan, and J. Bernussou. Continuous-time analysis, eigenstructure assignment, and H_2 synthesis with enhanced linear matrix inequalities (LMI) characterizations. IEEE Transactions on Automatic Control, 46(12):1941–1946, 2001.

[13] U. Shaked. Improved LMI representations for analysis and design of continuous-time systems with polytopic-type uncertainty. IEEE Transactions on Automatic Control, 46(5):652–656, 2001.

[14] M. Chilali and P. Gahinet. H_∞ design with pole placement constraints: An LMI approach. IEEE Transactions on Automatic Control, 41(3):358–367, 1996.

[15] C. Du, L. Xie, J. N. Teoh, and G. Guo. An improved mixed $\mathcal{H}_2/\mathcal{H}_\infty$ control design for hard disk drives. IEEE Transactions on Control Systems Technology, 13(5):832–839, 2005.

[16] M. Chilali, P. Gahinet, and P. Apkarian. Robust pole placement in LMI regions. IEEE Transactions on Automatic Control, 44(12):2257–2270, 1999.

[17] M. C. De Oliveira, J. Bernussou, and J. C. Geromel. A new discrete-time robust stability condition. Systems & Control Letters, 37(4):261–265, 1999.

[18] V. J. S. Leite and P. L. D. Peres. An improved LMI condition for robust *D*-stability of uncertain polytopic systems. IEEE Transactions on Automatic Control, 48(3):500–504, 2003.

[19] V. F. Montagner and P. L. D. Peres. $\mathcal{H}_\infty$ parameter-dependent state feedback control of linear time-varying systems in polytopic domains. In 44th IEEE Conference on Decision and Control, 2005 and 2005 European Control Conference (CDC-ECC'05), Seville, Spain, pp. 5006–5011, IEEE, 2005.

[20] D. Peaucelle, D. Arzelier, O. Bachelier, and J. Bernussou. A new robust *D*-stability condition for real convex polytopic uncertainty. Systems & Control Letters, 40(1):21–30, 2000.

[21] T. Ishihara, H.-J. Guo, and H. Takeda. A design of discrete-time integral controllers with computation delays via loop transfer recovery. Automatica, 28(3):599–603, 1992.

Chapter 2

Discrete-Time Sliding Mode Control with Outputs of Relative Degree More than One

Sohom Chakrabarty, Bijnan Bandyopadhyay and
Andrzej Bartoszewicz

Additional information is available at the end of the chapter

http://dx.doi.org/10.5772/intechopen.68931

Abstract

This work deals with sliding mode control of discrete-time systems where the outputs are defined or chosen to be of relative degrees more than one. The analysis brings forward important advancements in the direction of discrete-time sliding mode control, such as improved robustness and performance of the system. It is proved that the ultimate band about the sliding surface could be greatly reduced by the choice of higher relative degree outputs, thus increasing the robustness of the system. Moreover, finite-time stability in absence of uncertainties is proved for such a choice of higher relative degree output. In presence of uncertainties, the system states become finite time ultimately bounded in nature. The work presents in some detail the case with relative degree two outputs, deducing switching and non-switching reaching laws for the same, while for arbitrary relative degree outputs, it shows a general formalisation of a control structure specific for a certain type of linear systems.

Keywords: discrete time, sliding mode control, finite-time stability, robust control, ultimate band

1. Introduction

Sliding mode control is a robust control technique, which is able to make the system insensitive towards a particular class of uncertainties in finite time. Such uncertainties, known as matched uncertainties, are those that appear along the input channel of the system and can be nullified by a simple switching control structure when the disturbance is bounded in nature. The switch happens about a surface in the space of the state variables and is called a sliding or a switching surface. The sliding variable $s = s(x)$ denotes how far the system states are from the sliding surface $S = \{x : s(x) = 0\}$. The control brings the system monotonically towards the sliding surface, thus $|s(t)|$ reducing until it becomes zero at a finite time. This is called the reaching phase. Once the system hits the surface, it stays there for all future times, thus making the system dynamics independent of the matched uncertainties and dependent only on the sliding surface parameters. Chosen appropriately, one can ensure that the system states become at least asymptotically stable during this phase called sliding motion of the system [15].

However, in practice, this beautiful property of sliding mode control could not be realized because of physical limitations of an actuator. Theoretically, the control needs to switch about the sliding surface with infinite frequency in order to be insensitive towards bounded matched uncertainties, but no real actuators can offer switching with infinite frequency. This causes chattering, which are high frequency actuator action giving rise to unmodelled dynamics excitation in the system as well as rapid degradation of the physical system. Moreover, measurements by sensors and control computation in a digital computer take place in finite-time intervals in modern times, thus ripping off the properties of continuous sliding mode control which made it theoretically so appealing.

To remove this gap between theory and practice, researchers developed the theory of discrete-time sliding mode control (DSMC) in [1–3, 16, 17, 19, 20, 22, 23]. Moreover, there are many inherently discrete-time systems that appear in nature as well as in engineering. For such discrete representation of a system, it was shown that the states of these systems can no longer hit the sliding surface and stay there in presence of disturbances. The best that can be achieved is ultimate boundedness of the system about the sliding surface in finite time. Hence, robustness of the system gets defined by the width of this ultimate band for discrete-time systems. It then becomes imperative that research takes place in the direction to reduce the width of the ultimate band, ensuring better robustness of the system. The work in this chapter is motivated by this objective and in the sequel it is shown how the choice of the relative degree of the output (or the sliding variable) to be greater than one, positively influences the robustness as well as the performance of the system as defined above. From this point and further in the chapter, the terms 'output' and 'sliding variable' will be used interchangeably, as sliding variable can be viewed as a constructed output of the system.

Traditionally, DSMC has been developed by taking outputs of relative degree one, i.e. there is only unit delay between the output and the input of the system. This has given rise to proposals of various reaching laws of the form $s(k+1) = f(s(k))$, where $s(k)$ is the sliding variable at the kth time step. These reaching laws make $|s(k)|$ approach an ultimate band about the sliding surface in finite time. One can readily calculate the control that does so from the reaching law, since $s(k+1)$ contains the control $u(k)$, when calculated from the system model. The most well-known reaching laws are laid down in Refs. [2, 3, 17]. Of the above, the first two papers deal with non-switching reaching laws, whereas the third one had proposed a switching reaching law. Even to this day, reaching law propositions form an important area of work in discrete-time sliding mode control, with different reaching laws favouring the design of control for a particular type of system. Some of these reaching laws are found in Refs. [5–11, 21, 24, 25].

The unity relative degree assumed in all the above works is also their major limitation. While it is the normal case to consider, there is no real restriction on the choice of this relative degree. In some system structures, the output can be naturally of relative degree more than one. In http://dx.doi.org/10.5772/intechopen.68931

others, one can easily construct an output with higher relative degree and consider it to the sliding variable to go about the analysis. In the recent studies [13, 14] which constitute the content in this chapter, it is shown that when this apparent limitation is lifted, we get reduced width of ultimate band, thus increasing robustness, as well as finite-time stability during sliding in absence of uncertainties. The latter is an important achievement, as previously finite-time stability during sliding for discrete-time systems had not been achieved. Only in Ref. [18], such finite-time stability of states had been achieved during sliding, but with specific design of surface parameters. With relative degree more than one, this finite-time stability of the system states during sliding is always guaranteed for a wide range of choices of the surface parameters.

The chapter is written as follows: in Section 2, an idea on the relative degree of outputs for discrete-time systems is given, which is used in the theoretical developments in the remainder of the chapter. In Section 3, a detailed work with reaching law propositions is done for relative degree two outputs for general linear time-invariant (LTI) systems of order n. For arbitrary relative degree outputs, a generalized control structure is proposed for a specific form of LTI systems in Section 4, in which the relative degree r is equal to the order n of the system. Improved robustness and finite-time stability are proved for all cases in both the sections. Simulation examples are also shown in each section, which corroborate the theoretical developments. The chapter ends with discussing the main results and implications thereof.

2. Relative degree for discrete-time systems

The concept of relative degree is well understood for continuous-time systems. The definition can be written as follows:

Definition 1: For a continuous-time system

$$\dot{x} = f_c(t, x, u) \tag{1}$$

the output $y(t)$ is said to be of relative degree r if $y^r = g_r(t, x, u)$ and $y^i = g_i(t, x)\ \forall\ 0 \le i < r$, where $u(t)$ is the control input and y^p denotes the pth time derivative of y.

The above definition means that the control first appears physically in the rth derivative of the output $y(t)$ and not before that.

The concept of relative degree for discrete-time systems can be easily understood by making a parallel of the above definition in the discrete-time domain. The derivative operator in continuous time becomes the difference operator in discrete time. Each difference introduces a delay between the output and the input of the system. With this in mind, one can propose the definition of relative degree for discrete-time systems as follows:

Definition 2: For a discrete-time system

$$x(k+1) = f_d(k,\ x(k),\ u(k)) \tag{2}$$

the output $y(k)$ is said to be of relative degree r if $y(k+r) = h_r(k,\ x(k),\ u(k))$ and $y(k+i) = h_i(k,\ x(k))\ \forall\ 0 \leq i < r$, where $u(k)$ is the control input and $y(k+p)$ denotes the p unit delays of y.

Physically, the above definition means that the control first appears in the rth delay of the output $y(k)$ and not before that. For a simple LTI system (A, B, C), this will mean that $CA^{i-1}B = 0\ \forall\ i = 1\ to\ (r-1)$ and $CA^rB \neq 0$.

3. Systems with relative degree two output

Let us consider a discrete-time LTI system in the regular form as

$$\begin{aligned} x_1(k+1) &= A_{11}x_1(k) + A_{12}x_2(k) \\ x_2(k+1) &= A_{21}x_1(k) + A_{22}x_2(k) + B_2u(k) + B_2f(k) \end{aligned} \tag{3}$$

where $x_1(k) \in \mathbb{R}^{n-m}$ and $x_2(k) \in \mathbb{R}^m$ are the n states and $u(k) \in \mathbb{R}^m$ is the control input. The disturbance $f(k) \in \mathbb{R}^m$ is assumed to be bounded as $||f(k)|| \leq f_m$.

Obviously $A_{11} \in \mathbb{R}^{(n-m)\times(n-m)}$, $A_{12} \in \mathbb{R}^{(n-m)\times m}$, $A_{21} \in \mathbb{R}^{m\times(n-m)}$,$A_{22} \in \mathbb{R}^{m\times m}$ and $B_2 \in \mathbb{R}^{m\times m}$. Let us assume $\det(B_2) \neq 0$. Written in the standard form $x(k+1) = Ax(k) + B(u(k) + f(k))$ for LTI systems, we shall have $A = \begin{bmatrix} A_{11} & A_{12} \\ A_{21} & A_{22} \end{bmatrix}$ and $B = \begin{bmatrix} 0 \\ B_2 \end{bmatrix}$.

3.1. Asymptotic stability with relative degree one output

A relative degree one output for the discrete-time system as in Eq. (3) can be proposed as

$$s_1(k) = C_1^T x(k) = Cx_1(k) + I_m x_2(k) \tag{4}$$

where $C \in \mathbb{R}^{m\times(n-m)}$ and the suffix 1 denotes relative degree one. Then

$$C_1^T B = [\,C \quad I_m\,]\begin{bmatrix} 0 \\ B_2 \end{bmatrix} = B_2 \tag{5}$$

and we can calculate the control $u(k)$ from

$$s_1(k+1) = C_1^T Ax(k) + C_1^T Bu_1(k) + C_1^T Bf(k) \tag{6}$$

using some relative degree one reaching law for $s(k)$, since B_2 is non-singular.

Design of C is done considering closed-loop performance during sliding motion of the nominal system, i.e. system with $f(k) = 0$. When the system is sliding, output $s_1(k)$ is zero, which makes $x_2(k) = -Cx_1(k)$. Hence, the closed loop during sliding becomes

$$x_1(k+1) = (A_{11} - A_{12}C)x_1(k) \tag{7}$$

which is traditionally made asymptotically stable by choosing $\lambda_{\max}(A_{11} - A_{12}C) < 1$. Since $x_2(k)$ is algebraically related to $x_1(k)$, it also settles down to zero asymptotically.

3.2. Finite-time stability with relative degree two output

For the system in Eq. (3), a relative degree two output can be

$$s_2(k) = C_2^T x(k) = Cx_1(k) \tag{8}$$

where $C \in \mathbb{R}^{m\times(n-m)}$ can be chosen same as in Eq. (4) or different, but satisfying the conditions in Theorem 1 below. The suffix 2 is used to denote relative degree two.

Now $C_2^T B = [C \quad 0]\begin{bmatrix} 0 \\ B_2 \end{bmatrix} = 0$ clearly shows that

$$s_2(k+1) = C_2^T Ax(k) \tag{9}$$

as calculated from the system dynamics in Eq. (3) does not contain the control input $u(k)$. Then we need to further assume $C_2^T B = [C \quad 0]\begin{bmatrix} A_{11} & A_{12} \\ A_{21} & A_{22} \end{bmatrix}\begin{bmatrix} 0 \\ B_2 \end{bmatrix} = CA_{12}B_2$ to be non-singular so that the output $s_2(k)$ is of relative degree two. Then we obtain

$$s_2(k+2) = C_2^T A^2 x(k) + C_2^T AB(u_2(k) + f(k)) \tag{10}$$

by adding one more delay to Eq. (9). The control input $u(k)$ can now be obtained using Eq. (10).

Theorem 1. *If $Ker(C) = 0$ and $\det(CA_{12}) \neq 0$, then the output $s_2(k)$ with relative degree two as designed in Eq.* (8) *ensures finite-time stability of the states of the system in Eq.* (3) *during sliding, in absence of the disturbance $f(k)$.*

Proof. During sliding, $s_2(k) = Cx_1(k) = 0$. If $Ker(C) = 0$, it follows that $x_1(k) = 0$ during sliding. Also, we have $s_2(k+1) = CA_{11}x_1(k) + CA_{12}x_2(k) = 0$ during sliding which implies $x_2(k) = -(CA_{12})^{-1}CA_{11}x_1(k)$. As $x_1(k) = 0$, it follows that $x_2(k) = 0$ as well, since CA_{12} is assumed to be non-singular. Hence, all the states become zero at the same instant as the output hits zero. This happens in finite time for any appropriately designed reaching law, which can bring the nominal system to the sliding surface in finite time. Thus, one can conclude that the system states become finite-time stable with the choice of relative degree two output.

Note that, $Ker(C) = 0$ is only a sufficient condition and not a necessary one in order to achieve finite-time stability of system states. The above theorem points out an important achievement in the closed-loop reduced order dynamics compared to the choice of the relative degree one output. Of course, if there is a disturbance, then the finite-time stability would be changed to finite time-bounded stability, i.e. the system states will only enter an ultimate band in a finite time and stay there.

Remark 1. *In simulations, the parameter C is chosen the same for both relative degree one and two outputs for comparison purposes. However, selection of the parameter C for relative degree two output* *does not in any way require apriori design of the same parameter for a relative degree one output. The property of finite-time stability is inherent to the relative degree two output systems provided C is selected as per the conditions in Theorem 1, which are easy to satisfy.*

3.3. Non-switching reaching law

In Ref. [3], a reaching law for discrete-time systems is introduced as

$$s(k+1) = s_d(k+1) + d(k)$$
$$s_d(k) = \begin{cases} \dfrac{k^* - k}{k^*} s(0) \text{ for } k < k^* \\ 0 \text{ for } k \geq k* \end{cases} \tag{11}$$

and $d(k)$ is an uncertainty derived from the system uncertainty $f(k)$. It is evident that this reaching law makes the sliding variable $|s(k)| \leq d_m \ \forall \ k \geq k^*$, i.e. d_m is the ultimate band for the sliding variable $s(k)$, where the uncertainty $d(k)$ is bounded as $|d(k)| \leq d_m$.

3.3.1. Ultimate band for relative degree one output

It is evident that

$$s_1(k+1) = C_1^T x(k+1) = C_1^T Ax(k) + C_1^T B(u_1(k) + f(k)) \tag{12}$$

which requires $d(k) = d_1(k) = C_1^T Bf(k)$ in Eq. (11) so that the control

$$u_1(k) = -\left(C_1^T B\right)^{-1} \left[\left(C_1^T A\right)x(k) - s_d(k+1)\right] \tag{13}$$

does not contain any uncertain terms. This makes the bound of $d_1(k)$ for relative degree one outputs as

$$d_{1m} = \left|\left|C_1^T B\right|\right| f_m = \left|\left|B_2\right|\right| f_m \tag{14}$$

which is the ultimate band δ_1 as well.

3.3.2. Ultimate band for relative degree two output

It is already shown that $s_2(k+1)$ does not contain the control input as well as the matched disturbance, being a relative degree two output. Hence, we obtain

$$s_2(k+2) = C_2^T x(k+2) = C_2^T A^2 x(k) + C_2^T AB(u(k) + f(k)) \tag{15}$$

containing the control input and this requires to extend the reaching law in Eq. (11) to find $s_2(k+2)$. It is done by taking the nominal part of the reaching law (without $d(k)$) and adding an unit delay to find $s_2(k+2)$. Then we include $d_2(k)$ to take care of the matched disturbance. This gives the extended reaching law for relative degree two outputs as

$$s_2(k+2) = s_d(k+2) + d_2(k)$$
$$s_d(k) = \begin{cases} \frac{k^* - k}{k^*} s(0) \text{ for } k < k^* \\ 0 \text{ for } k \geq k^* \end{cases} \quad (16)$$

With $d_2(k) = C_2^T ABf(k)$ in Eq. (16), the control input

$$u_2(k) = -\left(C_2^T AB\right)^{-1}[(C_2^T A^2 x(k) - s_d(k+2)] \quad (17)$$

does not contain any uncertain terms. The bound of $d_2(k)$ in this case is

$$d_{2m} = \left||C_2^T AB|\right| f_m \leq \left||CA_{12}|\right| \left||B_2|\right| f_m = \left||CA_{12}|\right| d_{1m} \quad (18)$$

which is the ultimate band δ_2 as well.

Theorem 2. *If in addition to the conditions in Theorem 1, C also satisfies $\lambda_{\max}(CA_{12}) < 1$, then the ultimate band δ_2 for the relative degree two output with reaching law in Eq.* (16) *is lesser than the ultimate band δ_1 for the relative degree one output with reaching law in Eq.* (11), *irrespective of whether the parameter C is chosen same for both relative degree cases.*

Proof. The property is straightforward to see from Eq. (18).

3.4. Switching reaching law

In Ref. [17], Gao et al. proposed a switching reaching law for discrete time SMC systems, which has the form

$$s_1(k+1) = \alpha s_1(k) - \beta_1 \text{sign}(s_1(k)) + d_1(k) \quad (19)$$

where $\alpha \in (0,1)$ and $\beta_1 > d_{1m}$ are real constants, $d_1(k)$ is the uncertainty derived from the system uncertainty $f(k)$ and bounded as $|d_1(k)| < d_{1m}$. At present there are two ways to analyse Gao's reaching law, one provided in Ref. [4] and the other in Ref. [12]. In this work, the well-known analysis established in Ref. [4] is followed.

3.4.1. Ultimate band for relative degree one output

It is already shown that

$$s_1(k+1) = C_1^T x(k+1) = C_1^T Ax(k) + C_1^T B(u_1(k) + f(k)) \quad (20)$$

which requires $d_1(k) = C_1^T Bf(k)$ in Eq. (19) so that the control input

$$u_1(k) = -\left(C_1^T B\right)^{-1}[C_1^T Ax(k) - \alpha C_1^T x(k) + \beta_1 \text{sign}\left(C_1^T x(k)\right)] \quad (21)$$

does not contain uncertain terms. This makes the bound of $d_1(k)$ for relative degree one outputs as

$$d_{1m} = \left\|C_1^T B\right\| f_m = \left\|B_2\right\| f_m \tag{22}$$

which is the same as Eq. (14) in Section 3.3.1.

As per the analysis in Ref. [4] of the reaching law in Eq. (19), we need $\beta_1 > \frac{(1+\alpha)}{(1-\alpha)} d_{1m}$ for crossing-recrossing $s_1(k) = 0$ at each successive step after crossing it for the first time. The ultimate band is then calculated as

$$\delta_1 = \beta_1 + d_{1m} > \frac{2d_{1m}}{1-\alpha} \tag{23}$$

3.4.2. Ultimate band for relative degree two output

It is already shown that $s_2(k+1)$ does not contain the input. Hence, we calculate

$$s_2(k+2) = C_2^T x(k+2) = C_2^T A^2 x(k) + C_2^T AB(u_2(k) + f(k)) \tag{24}$$

where the control input appears. This requires one to also extend the reaching law in Eq. (11) to find $s_2(k+2)$. This is done by taking the nominal part of the reaching law (i.e. with $d(k) = 0$) and adding another unit delay to find $s_2(k+2)$. Then we include $d_2(k)$ to take care of the matched disturbance. This gives the extended reaching law as

$$s_2(k+2) = \alpha^2 s_2(k) - \alpha\beta_2 \text{sign}(s_2(k)) - \beta_2 \text{sign}(s_2(k+1)) + d_2(k) \tag{25}$$

With $d_2(k) = C_2^T ABf(k)$ in Eq. (25), the control

$$u_2(k) = -\left(C_2^T AB\right)^{-1}\left[\left(C_2^T A^2 - \alpha^2 C_2^T\right)x(k) + \alpha\beta_2 \text{sign}\left(C_2^T x(k)\right) + \beta_2 \text{sign}\left(C_2^T Ax(k)\right)\right] \tag{26}$$

becomes devoid of any uncertain terms. The bound of $d_2(k)$ in this case is

$$d_{2m} = \left\|C_2^T AB\right\| f_m \leq \left\|CA_{12}\right\| \left\|B_2\right\| f_m = \left\|CA_{12}\right\| d_{1m} \tag{27}$$

which is same as Eq. (18) in Section 3.3.2. The task now is to determine the ultimate band δ_2 and the conditions on β_2 that needs to be satisfied. These are evaluated keeping in mind the property of crossing-recrossing about $s_2(k) = 0$ as imposed in the original work in Ref. [17] for relative degree one output. For simplicity, we perform the analysis assuming $s_2(k) \in R$. For a higher-dimensional output $s_2(k)$, the same analysis shall hold for each element of the vector.

Let us consider the sliding variable $s_2(k)$ at two consecutive time instants. In other words, we take into account the values of both $s_2(k)$ and $s_2(k+1)$, where k is any non-negative integer. Then, one can either have $\text{sign}(s_2(k+1)) = \text{sign}(s_2(k))$ or $\text{sign}(s_2(k+1)) = -\text{sign}(s_2(k))$.

Lemma 1. *If* $\beta_2 > \frac{d_{2m}}{1+\alpha}$ *and* $\text{sign}(s_2(k+1)) = \text{sign}(s_2(k))$, *then* $|s_2(k+2)|$ *is strictly smaller than* $|s_2(k)|$ *or crosses the hyperplane* $s_2(k) = 0$.

Proof. For $\text{sign}(s_2(k+1)) = \text{sign}(s_2(k)) = 1$, from Eq. (25) we get

$$s_2(k+2) \le \alpha^2 s_2(k) - (1+\alpha)\beta_2 + d_{2m} < s_2(k) \tag{28}$$

since $\beta_2 > \frac{d_{2m}}{1+\alpha}$.

For $\text{sign}(s_2(k+1)) = \text{sign}(s_2(k)) = -1$, from Eq. (25) we get

$$s_2(k+2) \ge \alpha^2 s_2(k) + (1+\alpha)\beta_2 - d_{2m} > s_2(k) \tag{29}$$

It is straightforward to conclude from the above two inequalities that $|s_2(k+2)| < |s_2(k)|$ or $\text{sign}(s_2(k+2)) = -\text{sign}(s_2(k+1)) = -\text{sign}(s_2(k))$.

Lemma 1 can be geometrically interpreted as follows: if the states $x(k)$ and $x(k+1)$ are on the same side of the sliding hyperplane, then either $x(k+2)$ is at the same side of the hyperplane and closer to it than $x(k)$ or $x(k+2)$ is on the other side of the hyperplane.

As k is an arbitrary non-negative integer, the above lemma demonstrates that there exists such a finite $k_0 > 0$ that $\forall\, i < k_0$, we have $\text{sign}[s_2(i)] = \text{sign}[s_2(0)]$ and $\text{sign}[s_2(k_0)] = -\ \text{sign}[s_2(0)]$. That is, there exists a finite time instant k_0, at which the sliding variable $s_2(k)$ changes its sign. In other words, the system crosses the sliding surface in finite time.

Lemma 2. *If* $\beta_2 > \frac{d_{2m}}{1-\alpha}$ *and* $\text{sign}(s_2(k+1)) = -\ \text{sign}(s_2(k))$, *then* $\text{sign}(s_2(k+2)) = \text{sign}(s_2(k))$.

Proof. With $\text{sign}(s_2(k+1)) = -\ \text{sign}(s_2(k))$, from Eq. (25) we get

$$\begin{aligned} s_2(k+2) &= \alpha^2 s_2(k) - \alpha\beta_2 \text{sign}(s_2(k)) - \beta_2 \text{sign}(s_2(k+1)) + d_2(k) \\ &= \alpha^2 s_2(k) - \alpha\beta_2 \text{sign}\ (s_2(k)) + \beta_2 \text{sign}(s_2(k)) + d_2(k) \\ &= \alpha^2 s_2(k) + (1-\alpha)\beta_2 \text{sign}(s_2(k)) + d_2(k) \end{aligned} \tag{30}$$

Since $\beta_2 > \frac{d_{2m}}{1-\alpha}$, then for any $|d_2(k)| < d_{2m}$, we get $\text{sign}(s_2(k+2)) = \text{sign}(s_2(k))$.

As k is an arbitrary non-negative integer, the above lemma implies that $\beta_2 > \frac{d_{2m}}{1-\alpha}$ is both a necessary and sufficient condition for crossing-recrossing the sliding hyperplane $s_2(k) = 0$ at each successive step after crossing it for the first time. Furthermore, the condition on β_2 in Lemma 2 automatically guarantees that the condition on β_2 in Lemma 1 holds. This concludes that the former is a necessary and sufficient condition for generating the quasi-sliding mode in the sense of Gao [17]. Indeed, when $\beta_2 > \frac{d_{2m}}{1-\alpha}$ is satisfied, then the system crosses the sliding hyperplane in a finite time and then recrosses it again in every consecutive step. However, the sequence $\{|s(k)|\}$ may not necessarily approach zero monotonically, but the sequence of every alternate sample of $\{|s(k)|\}$ does. Ultimately, the quasi-sliding mode is achieved when $\{|s(k)|\}$ starts crossing-recrossing about $s(k) = 0$ at each time step.

With the help of these ideas, the ultimate band δ_2 for the sliding variable $s_2(k)$ can be found out, which gives a measure of the robustness of the system concerned. The ultimate band must be equal to the largest steady-state value of the sliding variable for the maximum disturbance $|d_2(k)| = d_{2m}$. This is obtained from Eq. (25) putting $s_2(k) = \delta_2$, which also gives the value of $s_2(k+2) = \delta_2$. Thus,

$$\delta_2 = \alpha^2\delta_2 - \alpha\beta_2 + \beta_2 + d_{2m} \tag{31}$$

which gives

$$\delta_2 = \frac{(1-\alpha)\beta_2 + d_{2m}}{(1-\alpha^2)} > \frac{2d_{2m}}{(1-\alpha^2)} \tag{32}$$

since $\beta_2 > \frac{d_{2m}}{(1-\alpha)}$.

Theorem 3. *If in addition to the conditions as in Theorem 1, C also satisfies $\sigma_{\max}(CA_{12}) < 1+\alpha$, then the ultimate band δ_2 for the relative degree two output with reaching law in Eq. (25) is lesser than the ultimate band δ_1 for the relative degree one output with reaching law in Eq. (19), irrespective of the parameter C chosen same for both relative degree cases.*

Proof. Let us consider $\rho > 1$. Then the inequalities in Eqs. (23) and (32) can be written as equalities multiplying the RHS with this ρ. This gives us

$$\delta_1 = \rho\frac{2d_{1m}}{(1-\alpha)} \quad \delta_2 = \rho\frac{2d_{2m}}{(1-\alpha^2)} \tag{33}$$

Taking into account the fact that $d_{2m} \leq ||CA_{12}||d_{1m}$, we get

$$\frac{\delta_2}{\delta_1} = \frac{2d_{2m}/2d_{1m}}{(1+\alpha)} \leq \frac{||CA_{12}||}{(1+\alpha)} \tag{34}$$

Hence, $\delta_2 < \delta_1$, if the condition $\lambda_{\max}(CA_{12}) < 1+\alpha$ is satisfied.

Here, ρ is selected the same for both the ultimate bands δ_1 and δ_2. It can be considered as a selection parameter for δ_1 which is kept same for the selection of δ_2 for fair comparison between the two ultimate bands.

Remark 2. *Compared to Theorem 2, the condition on C in Theorem 3 is more relaxed. Hence, with the switching reaching law in Eq. (25), we can decrease the ultimate band for relative degree two output with a less strict condition than required with the non-switching reaching law in Eq. (11).*

3.5. Simulation example

Simulation examples are shown for a second-order discrete LTI system with outputs of both relative degree one and two to compare performance.

We consider an inherently unstable dynamical system

$$x(k+1) = \begin{bmatrix} 1 & 1.2 \\ 5 & -1 \end{bmatrix} x(k) + \begin{bmatrix} 0 \\ 1 \end{bmatrix} (u(k) + f(k)) \tag{35}$$

where $f(k)$ is a disturbance assuming value $+0.1$ for the first half of the simulation cycle and -0.1 for the last half. The disturbance is chosen at these extremities to bring out the worst behaviour of the system. The comparison between choices of relative degree one and two outputs can be considered fair under such a scenario.

3.5.1. Non-switching reaching law

The reaching law of [3] with $k^* = 5$ is used for simulations. The surface parameter is selected as $C = 0.5$, which satisfies the conditions required in Theorem 2. The ultimate bands for the relative degree one and two outputs are calculated to be $\delta_1 = 0.1$ and $\delta_2 = 0.06$, respectively. **Figure 1** shows the plots of the output $s(k)$ along with a zoomed view to show the ultimate bands. The plots of the state variables and control input are given in **Figure 2**. The plots corresponding to relative degree one output are shown with a dotted line whereas those with relative degree two output are shown with a smooth line. It can be easily seen from **Figure 2**

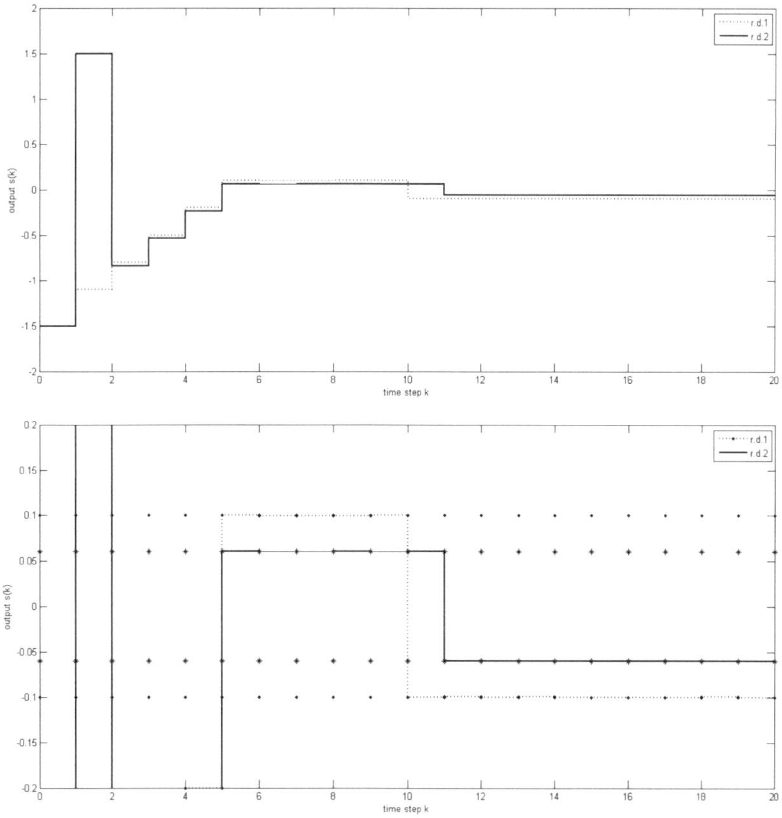

Figure 1. Sliding variable for non-switching reaching law.

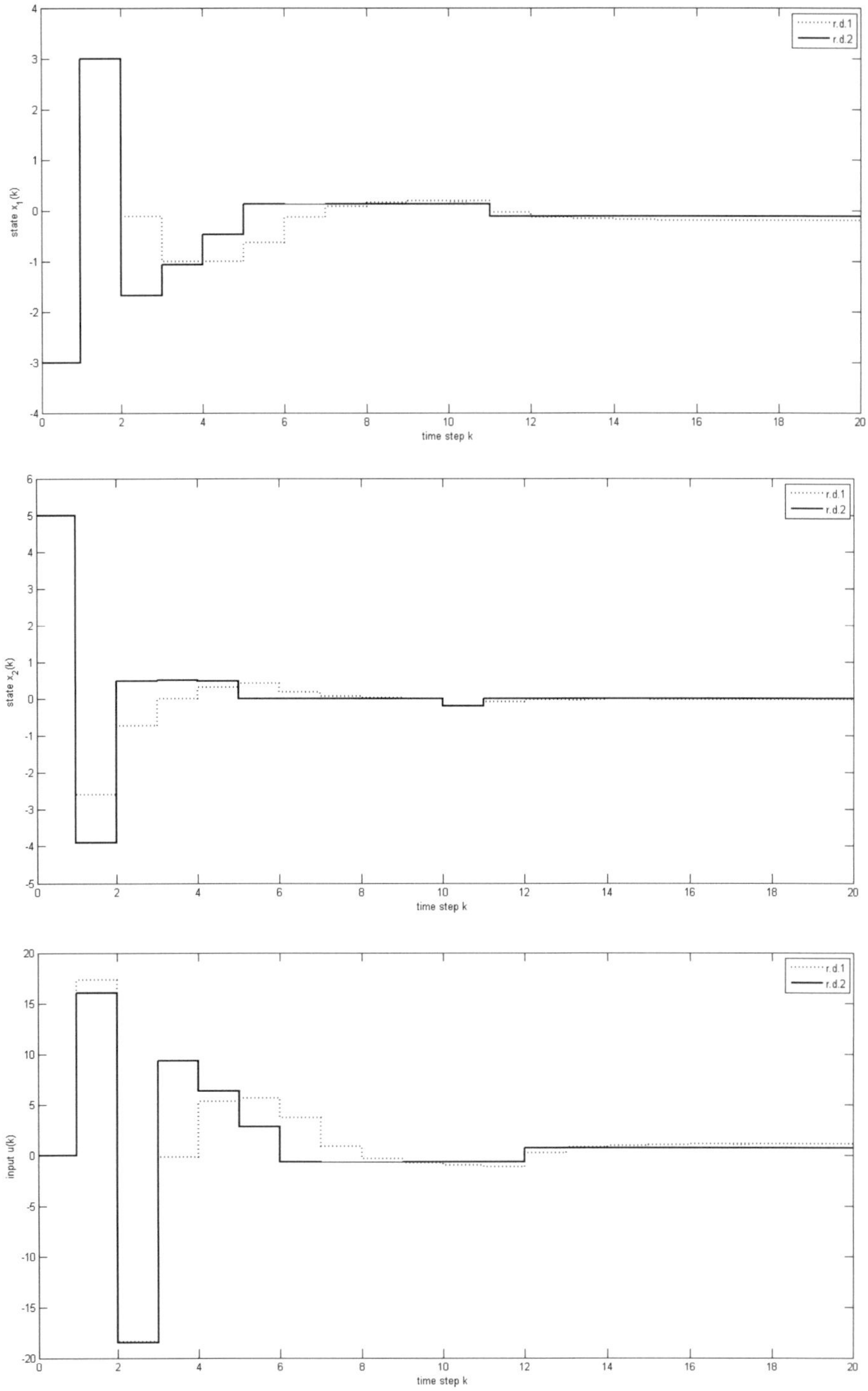

Figure 2. State variables and control input for non-switching reaching law.

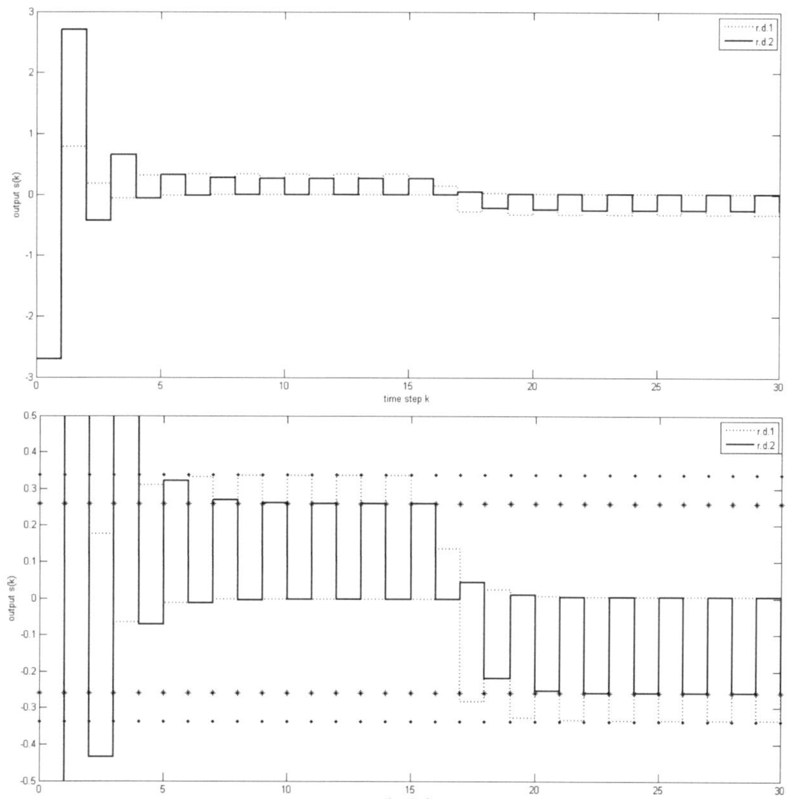

Figure 3. Sliding variable for switching reaching law.

that both the state errors as well as the control effort are also reduced for relative degree two output compared to relative degree one output.

3.5.2. Switching reaching law

The reaching law of Ref. [17] is used for simulations. The surface parameter is chosen as $C = 0.9$ which satisfies the conditions of Theorem 3 with $\alpha = 0.4$. For the purpose of simulations, $\rho = 1.01$ is selected which gives the ultimate bands as $\delta_1 = 0.3367$ and $\delta_2 = 0.2597$. For these values of the ultimate bands, $\beta_1 = 0.2367$ and $\beta_2 = 0.1836$ are calculated. **Figure 3** shows the plots of the output $s(k)$ along with a zoomed view to show the ultimate bands. The plots of the state variables and control input are given in **Figure 2**. The plots corresponding to relative degree one output are shown with a dotted line whereas those with relative degree two output are shown with a smooth line. It can be easily seen from **Figure 4** that both the state errors as well as the control effort are also reduced for relative degree two output compared to relative degree one output.

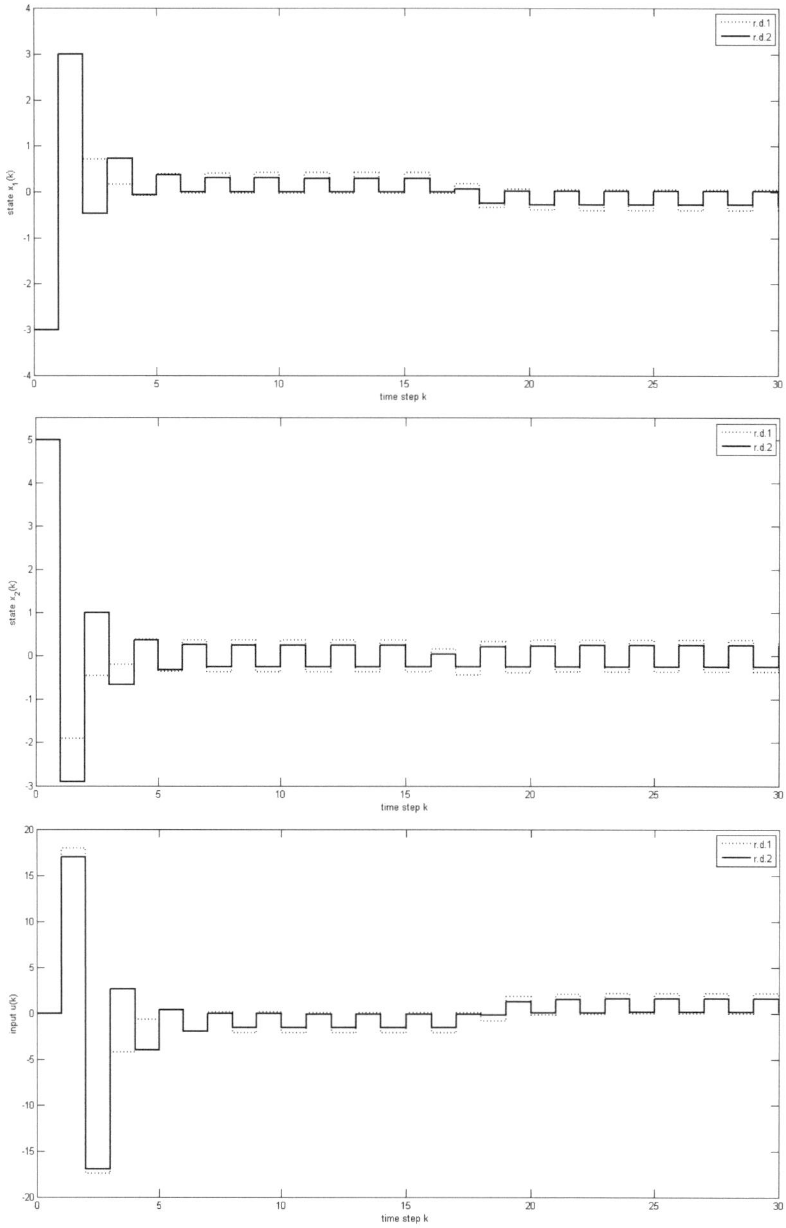

Figure 4. State variables and control input for switching reaching law.

4. Systems with arbitrary relative degree outputs

In Section 3, the system order n was arbitrary but the relative degree of the output was fixed to two. In this section, the relative degree is extended to arbitrary $r > 1$ where $r \in \mathbb{N}_{+}$. For the purpose of the theoretical development presented in this chapter, $r = n$ is considered, i.e. the relative degree of the output matches the system order. For such an assumption, the system structure can generally take a canonical form, called the lower Hessenberg form, whenever $r > 2$.

Consider a chain of n unit delays with the system output defined as $y(k) = cx_1(k)$, where $x_1(k)$ is the output of the last unit delay in the chain. Such a system structure is the popular controller canonical form for LTI systems, which can be obtained from any LTI system model by a simple linear transformation. However, with $r = n$, a model (A_n, B_n, C_n) of increased complexity can be considered, which is the lower Hessenberg form. This can be described by the system matrices $A_n = [a_{ij}]$, $i, j = 1$ to n, where $a_{ij} = 0 \, \forall \, i = 1$ to $(n-2)$, $j = (i+2)$ to n, $B_n = [0_{n-1} \; b]^T$ and $C_n = [c \;\; 0_{n-1}]$. Below is the general structure of the system matrix A_n:

$$A_n = \begin{bmatrix} a_{11} & a_{12} & 0 & 0 & \cdots & 0 & 0 \\ a_{21} & a_{22} & a_{23} & 0 & & 0 & 0 \\ & \vdots & & & \ddots & \vdots & \\ a_{(n-2)1} & a_{(n-2)2} & a_{(n-2)3} & a_{(n-2)4} & & a_{(n-2)(n-1)} & 0 \\ a_{(n-1)1} & a_{(n-1)2} & a_{(n-1)3} & a_{(n-1)4} & \cdots & a_{(n-1)(n-1)} & a_{(n-1)n} \\ a_{n1} & a_{n2} & a_{n3} & a_{n4} & & a_{n(n-1)} & a_{nn} \end{bmatrix}$$

Of course, $y(k)$, $u(k)$ and $f(k)$ are all scalar functions and the structure ensures that $y(k)$ is of relative degree $r = n$ as per the definition given in Section 2.

4.1. Finite-time stability of all states

Let us consider the system

$$x(k+1) = A_n x(k) + B_n(u(k) + f(k)) \; y(k+1) = C_n x(k) \tag{36}$$

with $f(k) = 0$. Assuming this nominal system reaches sliding mode, the following proposition can be made.

Theorem 4. *If the output of the system in Eq. (36) is of relative degree $r = n$, then $x_1(k) = x_2(k) = \ldots = x_n(k) = 0 \; \forall \; k \geq K$, where K is the time step at which the output $y(k)$ starts sliding, i.e. $y(k) = 0 \; \forall \; k \geq K$.*

Proof. During sliding, $y(k) = cx_1(k) = 0 \; \forall \; k \geq K$ implying $x_1(k) = 0 \; \forall \; k \geq K$ since $|c| \in (0, \infty)$. Now, obviously $y(k+1) = 0 \; \forall \; k \geq K$. This means

$$0 = ca_{11}x_1(k) + ca_{12}x_2(k) \; \forall \; k \geq K \tag{37}$$

implying $x_2(k) = 0 \; \forall \; k \geq K$ as $x_1(k) = 0 \; \forall \; k \geq K$ and $|a_{11}|$, $|a_{12}| \in (0, \infty)$ as per the system structure.

Similarly, $y(k+2) = y(k+3) = \ldots = y(k+n-1) = 0 \; \forall \; k \geq K$ and proceeding in the same line of argument, it can be shown that $x_3(k) = x_4(k) = \ldots = x_n(k) = 0 \; \forall \; k \geq K$. This implies that every state hits zero in finite time, which is the same as the time instant when the output hits zero, and stays there for all future times.

It is obvious that in the presence of uncertainty $f(k)$, the states will not reach zero but remain inside some ultimate band $\forall\ k \geq K$.

4.2. Improved robustness of the system

With relative degree of the output equal to the order of the system, better robustness can be obtained when compared to usual outputs of relative degree one, by satisfying certain sufficient conditions. The robustness is measured by the width of the ultimate band of the output or the sliding variable. For this, systems with outputs of relative degree two and three are first discussed and then the result is generalized for arbitrary relative degree outputs.

For a relative degree one output of an n-order system in Eq. (36), $C_n B_n = b$ if the sliding surface is linear, i.e. $C_n = [c\ c_2 \ldots 1]$. Hence, the control can always be computed from Utkin's reaching law [6]

$$y(k+1) = d_1(k) \tag{38}$$

with $|d_1(k)| \leq d_{1m} = C_n B_n f_m = b f_m$. This gives the control as

$$u(k) = -(C_n B_n)^{-1} C_n A_n x(k) \tag{39}$$

devoid of any uncertain terms, for any system dimension n.

4.2.1. *Relative degree two outputs*

With system order $n = 2$, the LTI system becomes

$$x(k+1) = \begin{bmatrix} a_{11} & a_{12} \\ a_{21} & a_{22} \end{bmatrix} x(k) + \begin{bmatrix} 0 \\ b \end{bmatrix} (u(k) + f(k)) \tag{40}$$

The output

$$y(k) = c x_1(k) = C_2 x(k) \tag{41}$$

is clearly of relative degree two, since $C_2 B_2 = 0$ and $C_2 A_2 B_2 \neq 0$. Hence, one needs

$$y(k+2) = C_2 A_2^2 x(k) + C_2 A_2 B_2 (u_2(k) + f(k)) \tag{42}$$

to obtain the equivalent control from the extended Utkin's reaching law for relative degree two outputs, which is easily obtained from Eq. (38) as

$$y(k+2) = d_2(k) \tag{43}$$

with $|d_2(k)| \leq d_{2m} = C_2 A_2 B_2 f_m = c a_{12} b f_m$. This makes the control

$$u_2(k) = -(C_2 A_2 B_2)^{-1} C_2 A_2^2 x(k) \tag{44}$$

devoid of any uncertain terms.

Obviously, the output $y(k)$ will be bounded inside the ultimate band $\delta_2 = d_{2m}\ \forall\ k \geq 2$. For the output with relative degree one, the ultimate band is simply $\delta_1 = d_{1m} = bf_m$. From the above, it is straightforward to put down the below theorem.

Theorem 5. *For the same LTI system in Eq. (40), the equivalent control will lead to a decrease in the width of the ultimate band with an output of relative degree two compared to an output of relative degree one if* $ca_{12} < 1$.

4.2.2. Relative degree three systems

With system order $n = 3$, the LTI system becomes

$$x(k+1) = \begin{bmatrix} a_{11} & a_{12} & 0 \\ a_{21} & a_{22} & a_{23} \\ a_{31} & a_{32} & a_{33} \end{bmatrix} x(k) + \begin{bmatrix} 0 \\ 0 \\ b \end{bmatrix} (u(k) + f(k)) \tag{45}$$

The output

$$y(k) = cx_1(k) = C_3 x(k) \tag{46}$$

is clearly of relative degree three, since $C_3B_3 = C_3A_3B_3 = 0$ and $C_3A_3^2B_3 \neq 0$. Hence, one needs

$$y(k+3) = C_3A_3^3x(k) + C_3A_3^2B_3(u_3(k) + f(k)) \tag{47}$$

to obtain the control from the extended Utkin's reaching law for relative degree three outputs. This is easily obtained from Eq. (38) as

$$y(k+3) = d_3(k) \tag{48}$$

with $|d_3(k)| \leq d_{3m} = C_3A_3^2B_3f_m = ca_{12}a_{23}bf_m$. This makes the control

$$u_3(k) = -\left(C_3A_3^2B_3\right)^{-1} C_3A_3^3x(k) \tag{49}$$

devoid of any uncertain terms.

Obviously, the output $y(k)$ will be bounded inside the ultimate band $\delta_3 = d_{3m}\ \forall\ k \geq 3$. For the output with relative degree one, the ultimate band is simply $\delta_1 = d_{1m} = bf_m$. From the above, it is straightforward to put down the below theorem.

Theorem 6. *For the same LTI system in Eq. (40), the equivalent control will lead to a decrease in the width of the ultimate band with an output of relative degree three compared to an output of relative degree one if* $ca_{12}a_{23} < 1$.

4.2.3. Systems with outputs of arbitrary relative degree

With relative degree of the output equal to the order of the system for an arbitrary $r = n$, the system is as given in Eq. (36) and $y(k + r)$ needs to be calculated from the output equation

$$y(k) = cx_1(k) = C_r x(k) \tag{50}$$

In the same way as in previous subsections, the control devoid of any uncertainty can be derived as

$$u_r(k) = -\left(C_r A_r^{r-1} B_r\right)^{-1} C_r A_r^r x(k) \tag{51}$$

from the extended Utkin's reaching law

$$y(k+r) = d_r(k) \tag{52}$$

where $|d_r(k)| \le d_{\mathrm{rm}} = c\prod_{i=2}^{r} a(i-1)_i b f_m$.

Obviously, the output will be bounded inside an ultimate band $\delta_r = d_{\mathrm{rm}}\ \forall\ k \ge r$. From the above, it is straightforward to put down the following theorem.

Theorem 7. *For the same LTI system in Eq. (36), the equivalent control will lead to a decrease in the width of the ultimate band with an output of relative degree $r = n$ compared to an output of relative degree one if $c\prod_{i=2}^{r} a(i-1)_i < 1$.*

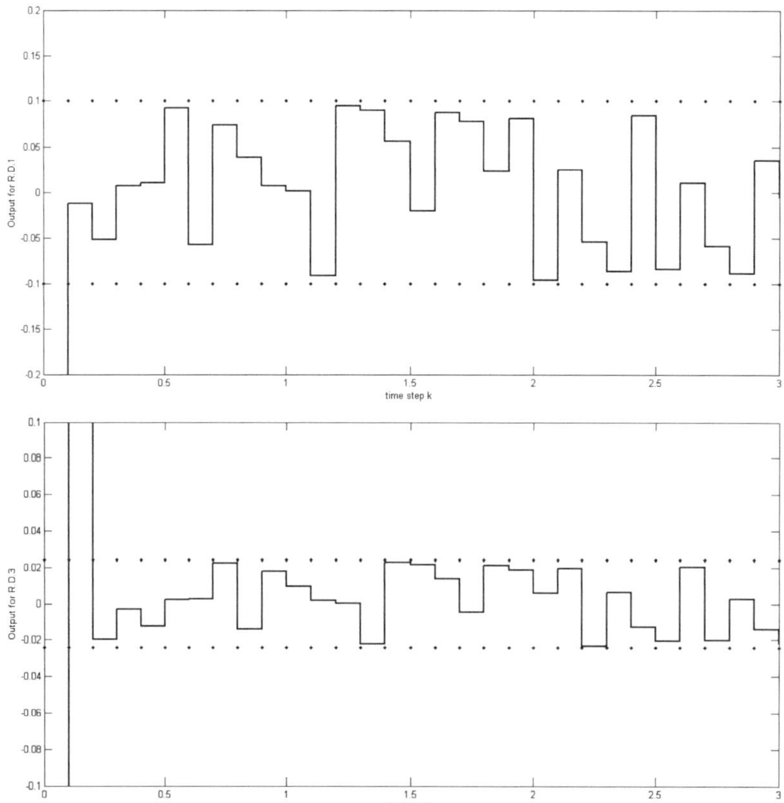

Figure 5. Comparing robustness of outputs with relative degree one and relative degree three.

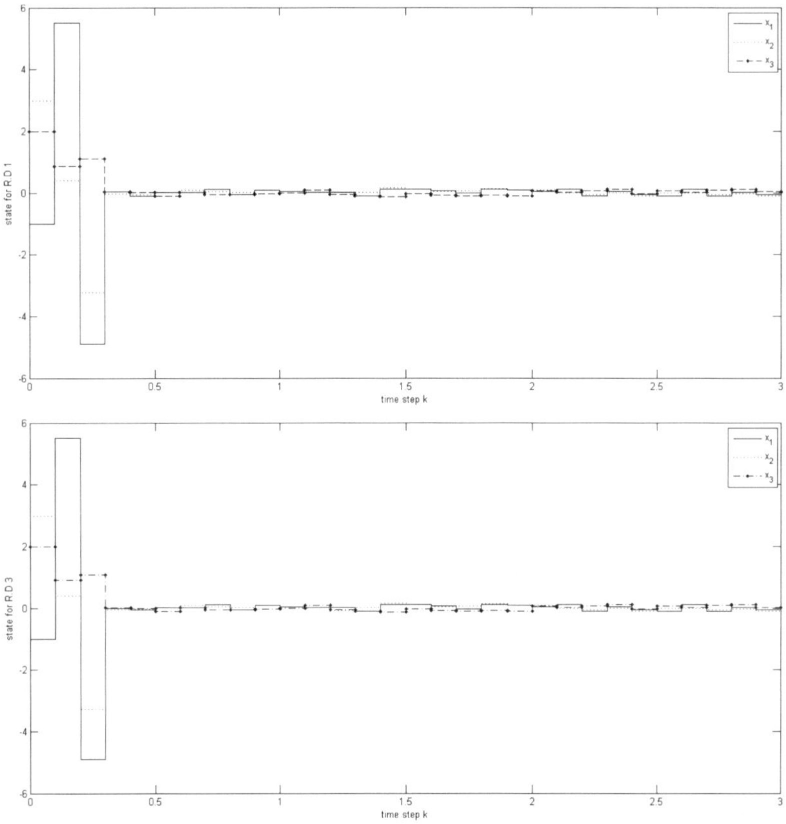

Figure 6. State dynamics for relative degree one and relative degree three with disturbance.

Remark 3. *In case of outputs with relative degree more than one, the scaling c can be dropped and simply* $y(k) = x_1(k)$*. Hence, the robustness entirely depends on the system parameters. It is thus possible that for some systems for which the parameters do not satisfy the condition in Theorem 7, the robustness worsens with choice of relative degree* $r = n$ *with Utkin's equivalent control law.*

4.3. Simulation result

A third-order discrete-time LTI system is considered with output of relative degree three for simulation. For comparison, the results for the output designed to be of relative degree one are also shown. It can be readily observed that with design parameters kept same for both, the system with relative degree three output shows better robustness in presence of disturbance and also achieves finite-time stability of all states in the absence of disturbance.

Let the system be

$$x(k+1) = \begin{bmatrix} -1 & 1.5 & 0 \\ -0.5 & 0.5 & -0.8 \\ -3 & 1 & 1 \end{bmatrix} x(k) + \begin{bmatrix} 0 \\ 0 \\ 1 \end{bmatrix} (u(k) + f(k)) \tag{53}$$

where $f(k)$ is a random number bounded by ± 0.1. The initial states are assumed to be $[-1\ 3 - 2]^T$.

An output of relative degree one is designed as

$$y_1(k) = [0.2 - 0.625 - 1]x(k) \tag{54}$$

which makes the poles of the reduced-order system in the sliding mode as 0.1 and −0.1, which are sufficiently nice pole placement to obtain asymptotic stability of the states fast enough.

The output of relative degree three is designed as

$$y_3(k) = [0.2\ 0\ 0]x(k) \tag{55}$$

by keeping the first entry of the output matrix same as in Eq. (54). The ultimate bands calculated for the relative degree one and three outputs are $\delta_1 = 0.1$ and $\delta_3 = 0.024$, respectively. The zoomed views of the outputs for the two cases are shown in **Figure 5**, with the ultimate band superimposed on each plot.

Figures 6 and **7** show the states and the control input for the two cases when the system is affected by the disturbance $f(k)$. Not much visible difference can be found between the

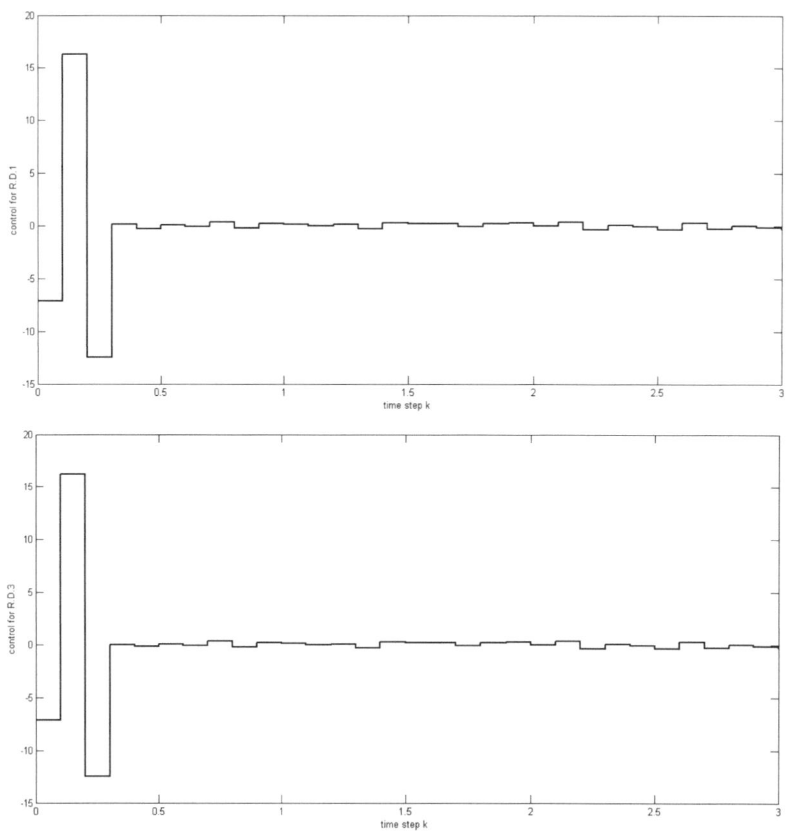

Figure 7. Control input for relative degree one and relative degree three with disturbance.

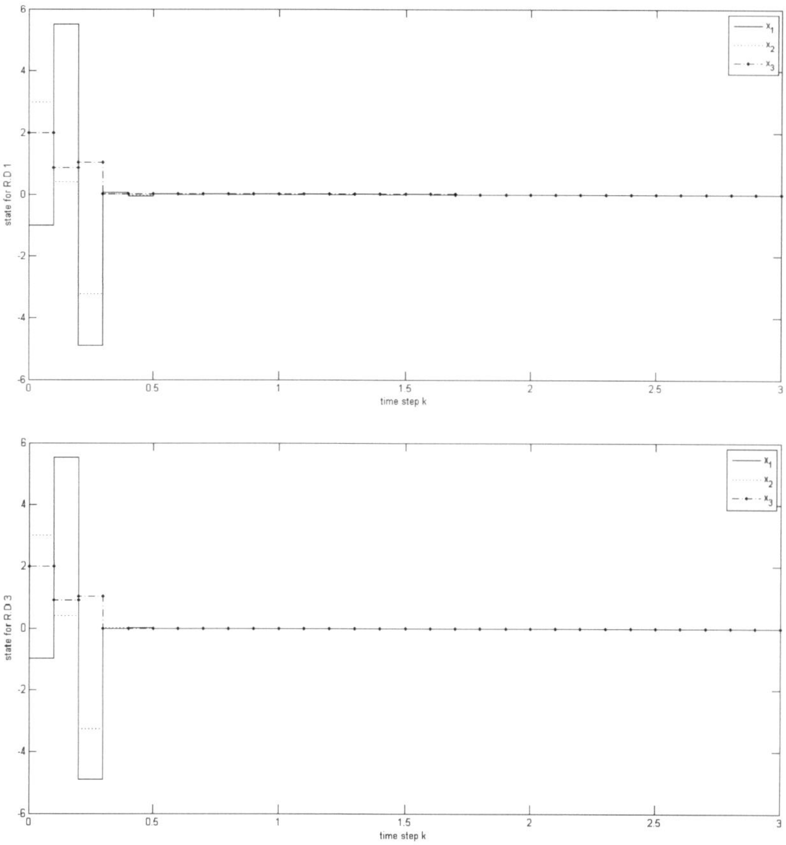

Figure 8. State dynamics for relative degree one and relative degree three without disturbance.

simulations of the states in **Figure 6** because of the presence of disturbance. However, in **Figure 8**, it is clear that the states of the system in absence of disturbance become finite-time stable for relative degree three output, whereas for relative degree one output, only asymptotic stability is achieved.

5. Conclusion

In this chapter, an important advancement in the direction of discrete-time sliding mode control is presented. As opposed to the traditional consideration of outputs of relative degree one, it is shown that with higher relative degree outputs, improved robustness and performance of the system can be guaranteed under certain conditions. New reaching laws are proposed for these higher relative degree outputs, which are extensions of existing reaching laws proposed in Refs. [2, 3, 17] for relative degree one outputs. These reaching laws are analysed to find out conditions for increased robustness of the system. Along with such increased robustness attributed to a reduction in the ultimate band of the sliding variable or output, the system states are also proved to be finite-time stable in absence of disturbance. In presence of disturbance, they are finite time ultimately bounded. Moreover, this finite time step is same as the time step at which the output hits the sliding surface.

Author details

Sohom Chakrabarty[1]*, Bijnan Bandyopadhyay[2] and Andrzej Bartoszewicz[3]

*Address all correspondence to: sohomfee@iitr.ac.in

1 Department of Electrical Engineering, Indian Institute of Technology Roorkee, Roorkee, India

2 IDP in Systems & Control, Indian Institute of Technology Bombay, Mumbai, India

3 Faculty of Electrical, Electronic, Computer and Control Engineering, Lodz University of Technology, Lodz, Poland

References

[1] Bandyopadhyay B, Janardhanan S. Discrete-time Sliding Mode Control: A Multirate Output Feedback Approach, LNCIS. 323rd ed. Springer-Verlag; Berlin Heidelberg, 2006.

[2] Bartolini G, Ferrara A, Utkin V. Adaptive sliding mode control in discrete-time systems. Automatica. 1995;**31**(5):769–773

[3] Bartoszewicz A. Discrete-time quasi-sliding-mode control strategies. IEEE Transactions on Industrial Electronics. 1998;**45**(4):633–637

[4] Bartoszewicz A. Remarks on discrete-time variable structure control systems. IEEE Transactions on Industrial Electronics. 1996;**43**(1):235–238

[5] Bartoszewicz A, Latosinski P. Discrete time sliding mode control with reduced switching: A new reaching law approach. International Journal of Robust and Nonlinear Control. 2016;**26**(1):47–68

[6] Bartoszewicz A, Lesniewski P. A new reaching law based sliding mode flow controller for connection-oriented data transmission networks. In: 13th IEEE Workshop on Variable Structure Systems; Jun 22–Jul 2; Nantes. 2014. pp. 1–6. Curran Associates, Inc., NY, USA.

[7] Bartoszewicz A, Lesniewski P. New switching and nonswitching type reaching laws for SMC of discrete time systems. IEEE Transactions on Control Systems Technology. 2016;**24**(2):670–677

[8] Bartoszewicz A, Lesniewski P. Reaching law approach to the sliding mode control of periodic review inventory systems. IEEE Transactions on Automation Science and Engineering. 2014;**11**(3):810–817

[9] Bartoszewicz A, Lesniewski P. Reaching law-based sliding mode congestion control for communication networks. IET Control Theory & Applications. 2014;**8**(17):1914–1920

[10] Chakrabarty S and Bandyopadhyay B. A generalized reaching law for discrete time sliding mode control. Automatica. 2015;**52**:83–86

[11] Chakrabarty S, Bandyopadhyay B. A generalized reaching law with different convergence rates. Automatica. 2016;**63**:34–37

[12] Chakrabarty S, Bandyopadhyay B. Minimum ultimate band design of discrete sliding mode control. Asian Journal of Control. 2015;**17**(5):1889–1897

[13] Chakrabarty S, Bandyopadhyay B, Moreno JA, Fridman L. Discrete sliding mode control for systems with arbitrary relative degree output. In: 14th International Workshop on Variable Structure Systems; 1–4 June 2016; Nanjing, China. 2016. pp. 160–165. Curran Associates, Inc., NY, USA.

[14] Chakrabarty S, Bartoszewicz A. Improved robustness and performance of discrete time sliding mode control systems. ISA Transactions. 2016;**65**:143–149

[15] Edwards C, Spurgeon SK. Sliding Mode Control: Theory and Applications. London: Taylor and Francis; 1998.

[16] Furuta K. Sliding mode control of a discrete system. Systems and Control Letters. 1990;**14**:145–152

[17] Gao W, Wang Y, Homaifa A. Discrete-time variable structure control systems. IEEE Transactions on Industrial Electronics. 1995;**42**(2):117–122

[18] Ignaciuk P, Bartoszewicz A. Sliding mode dead-beat control of perishable inventory systems with multiple suppliers. IEEE Transactions on Automation Science and Engineering. 2012;**9**(2):418–423

[19] Janardhanan S, Bandyopadhyay B. Multirate output feedback based robust quasi-sliding mode control of discrete-time systems. IEEE Transactions on Automatic Control. 2007;**52**(3):499–503

[20] Milosavljevic D. General conditions for the existence of a quasi-sliding mode on the switching hyperplane in discrete variable structure systems. Automation and Remote Control. 1985;**46**:307–314

[21] Niu Y, Ho DWC, Wang Z. Improved sliding mode control for discrete-time systems via reaching law. IET Control Theory and Applications. 2010;**4**(11):2245–2251

[22] Sarpturk SZ, Istefanopulos Y, Kaynak O. On the stability of discrete-time sliding mode control systems. IEEE Transactions on Automatic Control. 1987;**32**(10):930–932

[23] Utkin VI, Drakunov SV. On discrete-time sliding mode control. In: Proceedings of IFAC Symposium on Nonlinear Control Systems (NOLCOS); 14–16 July 1989; Capri, Italy. 1989. pp. 484–489

[24] Yuan L, Shen J, Xiao F, Wang H. A novel reaching law approach of quasi-sliding mode control for uncertain discrete-time systems. Journal of Central South University. 2012;**19** (9):2514–2519

[25] Zhu Q, Wang T, Jiang M, Wang Y. A new design scheme for discrete-time variable structure control systems. In: International Conference on Mechatronics and Automation; 9–12 August; 2009; Changchun, China, 2009. pp. 3475–3479. Curran Associates, Inc., NY, USA.

Chapter 3

Sliding Mode Control (SMC) of Image-Based Visual Servoing for a 6DOF Manipulator

Shutong Li, Ahmad Ghasemi, Wen-Fang Xie and Yanbin Gao

Additional information is available at the end of the chapter

http://dx.doi.org/10.5772/67521

Abstract

The accuracy and stability are two fundamental concerns of the visual servoing control system. This chapter presents a sliding mode controller for image-based visual servoing (IBVS) which can increase the accuracy of 6DOF robotic system with guaranteed stability. The proposed controller combines proportional derivative (PD) control with sliding mode control (SMC) for a 6DOF manipulator. Compared with conventional proportional or SMC controller, this approach owns faster convergence and better disturbance rejection ability. Both simulation and experimental results show that the proposed controller can increase the accuracy and robustness of a 6DOF robotic system.

Keywords: sliding mode control, image-based visual servoing, 6DOF robotic manipulator

1. Introduction

With the development of industrial manufacture technology, manufacture process is going forward, more dexterous and more efficient machines is needed to meet this large changes.

Visual servoing system comes into being; it can handle the dynamic interaction between the manipulator and environment and has been applied in various surroundings where high accuracy and strong robustness are needed, such as cell injection in Ref. [1], car steering, aircraft landing and missile tracking in Ref. [2]. Generally speaking, the system where the camera is used as a visual sensor in the feedback is referred to as visual servoing system. Depending on the configuration of the camera with respect to the robot, the visual servoing configuration can be classified as eye-in-hand where the camera is installed at the end effector and eye-to-hand where the camera is fixed in workspace [3]. In this chapter, more attention is focused on eye-in-hand configuration in visual servoing system.

Furthermore, visual servoing can be classified into three different classes: image-based visual servoing (IBVS), position-based visual servoing (PBVS) and hybrid visual servoing (HVS). Their performances have been precisely described in Refs. [2–6]. In comparison, although PBVS is convenient in actual application, a calibrated camera and a known geometric model of the target are needed. The control performance depends on the accuracy of the camera calibration and the object geometric model, which is difficult to ensure. IBVS directly uses image feature errors to calculate the control signal, which reduces computational delay and becomes less sensitive to the calibration and model errors in Refs. [3, 5]. In this chapter, more attention is focused on IBVS.

Various control methods have been presented to apply to IBVS system, including proportional-integral-derivative control (PID) in Refs. [4, 7, 8], predictive control in Ref. [9], sliding mode control (SMC) in Ref. [10], adaptive control in Ref. [11], and so on. The core of this control method is to generate a velocity vector or an acceleration vector as the control input to guide the end effector to the desired position, to complete the total control task.

Particularly, PID control has a wide range of applications because of its simple form and popularity among engineers. Some researchers have conducted various analyses on using proportional or proportional derivative (PD) controller to produce a velocity command in Ref. [7]; convergent property of proportional or PD controller is satisfactory, but sometimes sudden variation or small shakiness due to image noise or motion vibration will be caused. In order to address these issues, the researchers proposed the control scheme using PD controller and producing an acceleration command as the control input in Ref. [8]. The proposed method can solve the above-mentioned problems; however, only velocity signal can be accepted as control input in most visual servoing systems. SMC was considered to be successfully applied in some automatic control fields due to its insensitivity to model uncertainties and disturbance in Ref. [12]. Using SMC in IBVS or PBVS or robotic manipulator to guarantee the system robustness and good tracking performance has been reported in some literature in the recent years in Refs. [13–15]. Meanwhile, the chatting phenomenon of SMC also needs to be considered in some special situations.

In this chapter, a new enhanced IBVS scheme that combines PD control with SMC is proposed to generate the velocity profile to control the robotic manipulator. This PD-SMC method takes the advantages of PD and SMC methods. The stability of the enhanced IBVS method is proved by using Lyapunov method. It can achieve the better convergence performance, ensure the stability of the system and own the strong robustness when the system is subjected to uncertainty and noises.

This chapter is structured as follows. The visual servoing system model is described in Section 2. The enhanced controller is designed in Section 3. The system stability is analysed in Section 4. The simulations are performed in Section 5. The experiments are performed in Section 6. The concluding remarks and future work are mentioned in Section 7.

2. System description

In the IBVS system, the control problem can be expressed by obtaining the relation between the derivative of the image features and the camera spatial velocity in Refs. [3, 4]. The system model, which consists of a 6DOF manipulator with a camera mounted on its end effector, is shown in **Figure 1**.

Before going into the detailed discussion of the system model, the following notations are introduced. The camera spatial velocity can be noted by $u = (v_c, \omega_c)$, $v_c = (v_{cx}, v_{cy}, v_{cz})$ and $\omega_c = (\omega_{cx}, \omega_{cy}, \omega_{cz})$, which are the camera's linear velocity and angular velocity in Ref. [5]. Set the focal length of camera as f. A world point **P** in the camera frame is denoted by $P = (X, Y, Z)$, the projected coordinate in image space is denoted by $p = (x, y)$.

Using the velocity of the point relative to the camera frame, we can describe the relationship between the feature velocity and the camera velocity in normalized image coordinate in Ref. [3] as follows:

$$\begin{bmatrix} \dot{x} \\ \dot{y} \end{bmatrix} = \begin{bmatrix} -\frac{f}{Z} & 0 & \frac{x}{Z} & \frac{xy}{f} & -\frac{1+x^2}{f} & y \\ 0 & -\frac{f}{Z} & \frac{y}{Z} & \frac{1+y^2}{x} & -\frac{xy}{f} & x \end{bmatrix} \begin{bmatrix} v_{cx} \\ v_{cy} \\ v_{cz} \\ \omega_{cx} \\ \omega_{cy} \\ \omega_{cz} \end{bmatrix} \tag{1}$$

Due to a 6DOF manipulator that needs to be controlled, at least three feature points are necessary to avoid the interaction matrix singularities and the multiple global minima in Refs. [4, 8]. Nevertheless, three points may be required for interaction matrix singularities and the multiple global minima. For this reason, we use four feature points to control 6DOF in the space, the expression is written as follows:

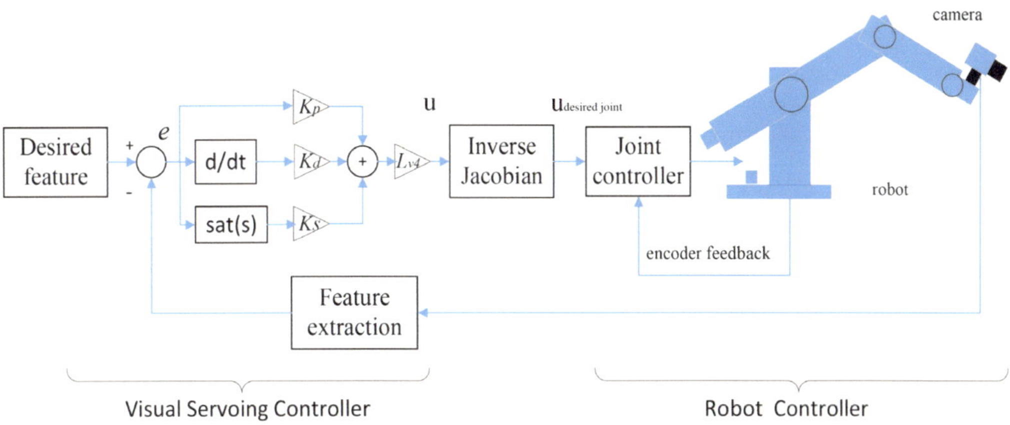

Figure 1. IBVS with eye-in-hand configuration.

$$\dot{e} = L_{v4}u \tag{2}$$

where L_{v4} is the interaction matrix.

$$L_{v4} = \begin{bmatrix} L_{v/p=p1} \\ L_{v/p=p2} \\ L_{v/p=p3} \\ L_{v/p=p4} \end{bmatrix} \tag{3}$$

$p = p1, \ldots, p4$ is the image feature points and e is the feature error. Since the image interaction matrix largely depends on the depth Z and camera intrinsic parameters such as focal length f, there exists some uncertainties in these parameters. In this chapter, we focus on dealing with the uncertainties on the depth. The range of the depth of the visual servoing system is assumed as $Z_{\min} \leq Z \leq Z_{\max}$. The estimated interaction matrix $\hat{L}_{v4}$ is used in the viusal servoing control design.

3. Controller design

The general design approach of a visual servoing controller is to use proportional control to generate the control signal. This approach is also applied to the conventional IBVS, the form can be described as follows:

$$u = -K\hat{L}_{v4}^{+}\dot{e}(t) \tag{4}$$

where $\hat{L}_{v4}^{+}$ is the pseudo inverse of the estimated interaction matrix, K is a positive definite matrix.

The proportional control is a prompt and timely control method. However, this method cannot eliminate the system residual error. In this chapter, PD control is used to replace the proportional control, which can improve the control quality with faster control convergence speed and smaller error. Meanwhile, in order to improve the system stability, the sliding model control is also adopted to compensate uncertainties of the system. This is an enhanced approach, which combines PD control with SMC base on IBVS, and can be called as hybrid PD-SMC method.

We define the sliding surface s, which will converge to 0 when the image feature errors go forward and stay on it all the time in Ref. [12].

$$s = e = \epsilon - \epsilon_d \tag{5}$$

where ϵ is the image plane feature and ϵ_d is the desired value of the feature. The basic visual servoing controller of IBVS is designed based on the above proportional control equation in Ref. [3], and it is described as the following first-order system:

http://dx.doi.org/10.5772/67521

$$\dot{e} + K_p e = 0 \tag{6}$$

Adding the sliding mode control, and applying PD control to the visual servoing system in Ref. [15], the modified control law should be considered as follows:

$$u = \hat{L}_{v4}^{+}\Big(- K_d\dot{e}(t) - K_p e(t) - K_s \mathrm{sgn}(s)\Big) \tag{7}$$

where K_s is a positive definite matrix and sgn(Â·) is the signum function.

Consider the above control scheme easily to have chatting phenomenon. In order to smooth the chattering, a saturation function is used to replace the sign function, and the control law is described as follows:

$$u = \hat{L}_{v4}^{+}\Big(- K_d\dot{e}(t) - K_p e(t) - K_s \mathrm{sat}(s)\Big) \tag{8}$$

where sat(·) is the saturation function, which is defined as follows:

$$\mathrm{sat}(s) = \begin{cases} s \text{ if } |s| \le 1 \\ \mathrm{sgn}(s) \text{ otherwise} \end{cases} \tag{9}$$

This control law is an enhanced IBVS scheme, which combines PD control with SMC together. SMC is well known for its robustness in Refs. [14–16]. By applying this control method, it is expected that this controller will achieve better robustness, faster convergence speed and higher accuracy. This will be demonstrated in both simulation and experiment sections.

4. Stability analysis

The stability analysis of the proposed controller is based on Lyapunov direct method in Ref. [12]. Consider the uncertainties in depth, the estimated interaction matrix can be described as follows:

$$(I + \Delta_{\min}) \le L_{v4}\hat{L}_{v4}^{+} \le (I + \Delta_{\max}) \tag{10}$$

where $\Delta_{\min}$ is a matrix of the uncertainties associated with lower bounds of estimated depth $Z_{\min}$ and $\Delta_{\max}$ is a matrix of the uncertainties associated with the upper bounds of the estimated depth $Z_{\max}$. A Lyapunov function is constructed as follows:

$$V = \frac{1}{2}s^T s \tag{11}$$

The time derivative of the Lyapunov function is obtained as follows:

$$\dot{V} = s^T \dot{s} \tag{12}$$

By substituting Eq. (8) into Eq. (2), the system error dynamic equation is obtained as follows:

$$\dot{e} = L_{v4}\hat{L}_{v4}^{+}\Big(- K_d\dot{e}(t) - K_p e(t) - K_s \mathrm{sat}(s)\Big) - \epsilon_d \tag{13}$$

Moving the term associated with $\dot{e}$ to the left of the equation yields

$$\dot{e} = L_{v4}\hat{L}_{v4}^{+}(I + K_d L_{v4}\hat{L}_{v4}^{+})^{-1}\Big(- K_p e - K_s \mathrm{sat}(s)\Big) - (I + K_d L_{v4}\hat{L}_{v4}^{+})^{-1}\epsilon_d \tag{14}$$

The time derivative of Lyapunov function is obtained as follows:

$$\dot{V} = L_{v4}\hat{L}_{v4}^{+}(I + K_d L_{v4}\hat{L}_{v4}^{+})^{-1}(-e^T K_p e - K_s|s|) - (I + K_d L_{v4}\hat{L}_{v4}^{+})^{-1} s^T \epsilon_d \tag{15}$$

It is noted that the rank of $L_{v4}\hat{L}_{v4}^{+}$ is 6, $L_{v4}\hat{L}_{v4}^{+}$ has two null vectors that satisfy $L_{v4}\hat{L}_{v4}^{+}x = 0$. It is know that $L_{v4}\hat{L}_{v4}^{+}$ has two null vectors that satisfy $\{L_{v4}\hat{L}_{v4}^{+} = 0,\ x \in \mathrm{R}^8,\ x \neq 0\}$. Assuming that x does not belong to the null space of $L_{v4}\hat{L}_{v4}^{+}$ in Refs. [4, 8], $L_{v4}\hat{L}_{v4}^{+}x > 0$ can be obtained. If K_d is chosen as a positive definite matrix,

$$K_d > 0 \tag{16}$$

The following formula can be ensured:

$$(I + K_d L_{v4}\hat{L}_{v4}^{+})^{-1} L_{v4}\hat{L}_{v4}^{+}x > 0 \tag{17}$$

K_s is chosen as follows:

$$K_s > \frac{\lambda_{\max}(I + \Delta_{\max})}{\lambda_{\min}(I + \Delta_{\min})^{-1}}\epsilon_d + \eta \tag{18}$$

where η is a diagonal positive definite matrix whose elements determine the decay rate of V to zero. $\lambda_{\min}$ and $\lambda_{\max}$ are the minima and maximum parameters, respectively.

According to the above conditions, the time derivative of Lyapunov function can be described as follows:

$$\dot{V} < L_{v4}\hat{L}_{v4}^{+}(I + K_d L_{v4}\hat{L}_{v4}^{+})^{-1}(-e^T K_p e - \eta|s|) < 0 \tag{19}$$

By applying Barbalat's lemma, it is obvious that $\dot{V} = 0$ when $t \to \infty$, the image feature error $e(t) \to \infty$. The stability of IBVS system is ensured.

5. Simulations

Simulations have been conducted on a 6DOF Puma 560 robot model by using MATLAB Robotics Toolbox and Machine Vision Toolbox in Ref. [3]. The 6DOF arm is chosen as the manipulator and the camera is mounted on the end effector, which assumes no transformation between the end effector and the camera. The camera characteristics are shown in **Table 1**. The maximum linear velocity of Puma 560 is 0.5 m/s according to the robot user manual in Refs. [17, 18].

To analyse and compare the performance of hybrid PD-SMC IBVS with the conventional IBVS, three simulation tests have been conducted, including pure translation and pure rotation of features, and disturbance rejection test. Four feature points are used in visual servoing control. The initial and desired positions of the image features are given in **Table 2**.

Test 1, in this test, a normal translational motion, is completed. **Figures 2** and **3** show the feature position error and joint velocity convergence situation of IBVS and enhanced IBVS (PD-SMC), respectively, under the pure translation condition. **Figure 4** shows the feature trajectory in image space under the pure translation condition.

Test 2, in this test, a pure rotational motion, is concluded. **Figures 5** and **6** show the feature position error and joint velocity convergence situation of IBVS and enhanced IBVS (PD-SMC), respectively, under the pure rotation condition. **Figure 7** shows the feature trajectory in image space under the pure rotation condition.

Test 3, in this test, a chirp signal as a disturbance, is added to demonstrate the robustness of the enhanced IBVS. **Figures 8** and **9** show the feature position error and joint velocity convergence situation of IBVS and enhanced IBVS (PD-SMC), respectively, under the disturbance. **Figure 7** shows the feature trajectory in image space under the disturbance.

Parameters	**Values**
Focal length	0.008 (m)
Principal point	(512, 512)
Camera resolution	1024 × 1024

Table 1. Camera parameters in simulations.

	Positions			
	$(x_1\ y_1)$	$(x_2\ y_2)$	$(x_3\ y_3)$	$(x_4\ y_4)$
Tests 1 and 3				
Initial	(360 401)	(361 611)	(570 610)	(573 402)
Desired	(412 412)	(412 612)	(612 612)	(612 412)
Test 2				
Initial	(360 401)	(361 611)	(570 610)	(573 402)
Desired	(362 506)	(466 612)	(572 506)	(466 403)

Table 2. Initial and desired positions.

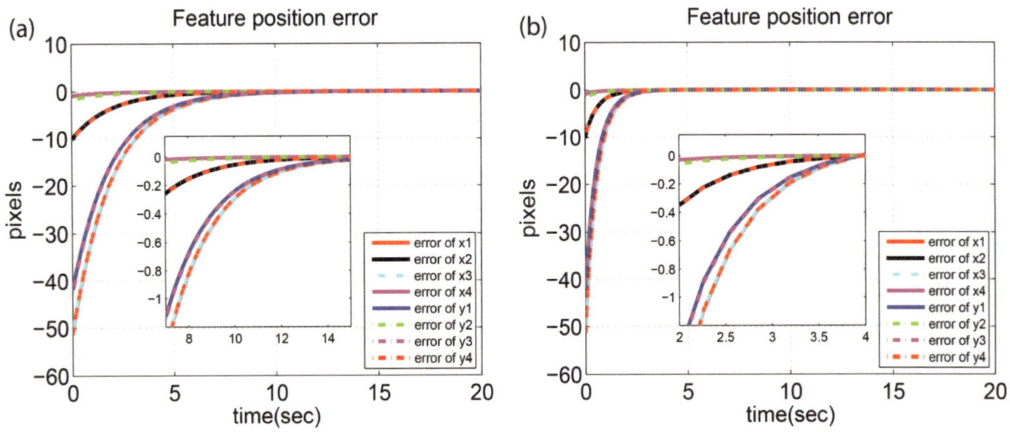

Figure 2. Feature error variation in pure translation test: (a) IBVS and (b) enhanced IBVS (PD-SMC).

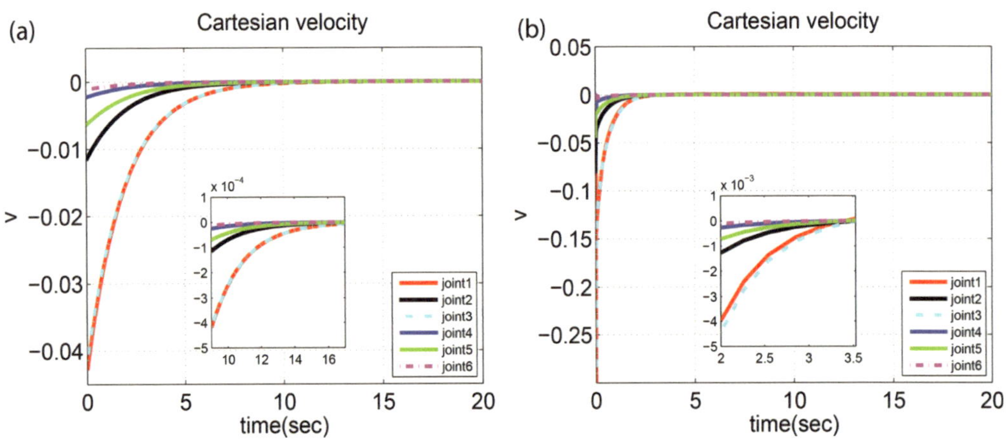

Figure 3. Joint velocity variation in pure translation test. (a) IBVS and (b) enhanced IBVS (PD-SMC).

According to three test results, it is obvious that the performance of enhanced IBVS is better than that of IBVS. More specifically, the simulation results demonstrate that the PD-SMC control system owns higher convergence rate, more accurate convergence state and strong robustness.

To further compare the performance of IBVS and enhanced IBVS, the performance index ISE (integrate square error) is adopted, which is defined as

$$\mathrm{ISE} = \int_{0}^{T} e^2(t)dt \tag{20}$$

The results are summarized in **Table 3**, where the 'ISE Total' represents the total integrate square error of feature error *x1,x2,x3,x4* and feature error *y1,y2,y3,y4*. It shows that the ISE of enhanced IBVS is smaller than that of IBVS in three tests.

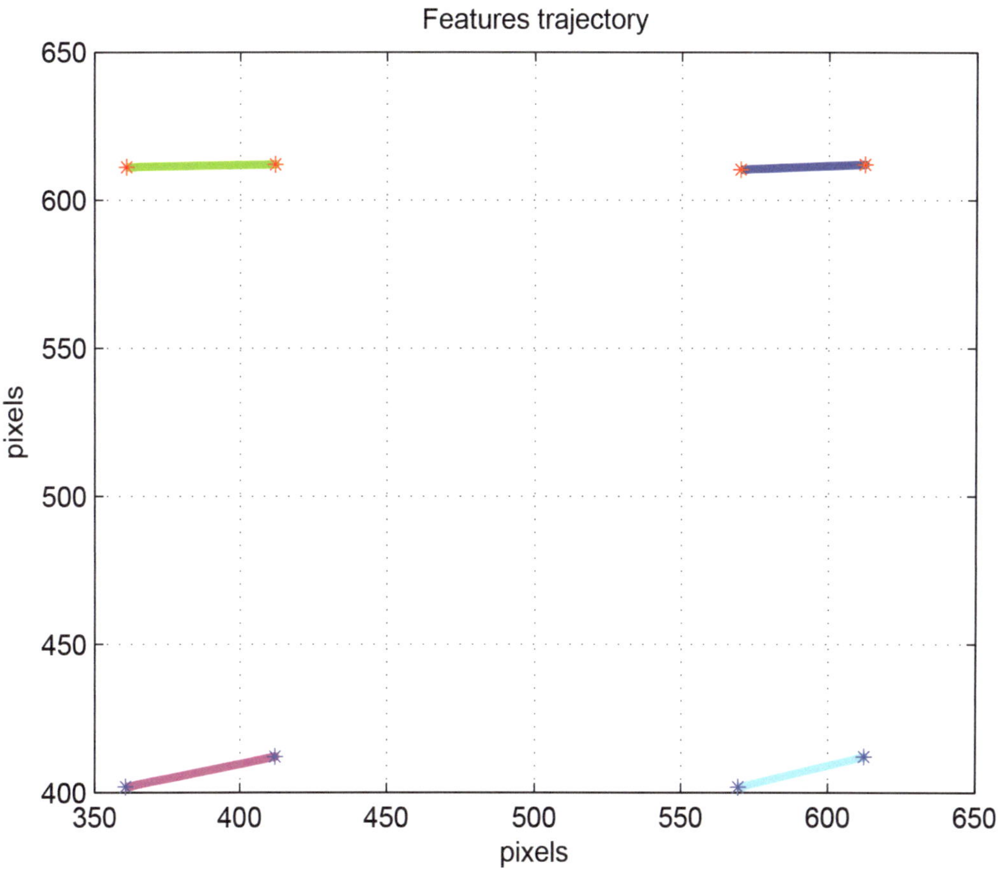

Figure 4. Feature trajectory in image space in Test 1.

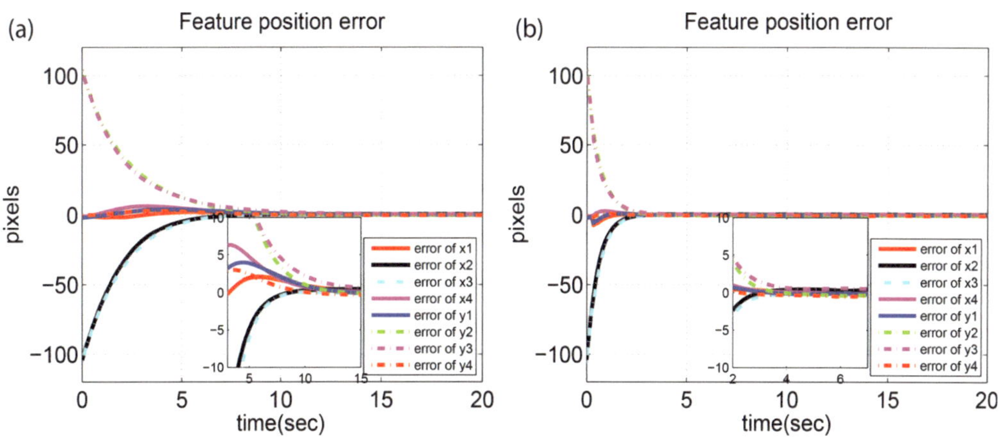

Figure 5. Feature error variation in pure rotation test. (a) IBVS and (b) enhanced IBVS (PD-SMC).

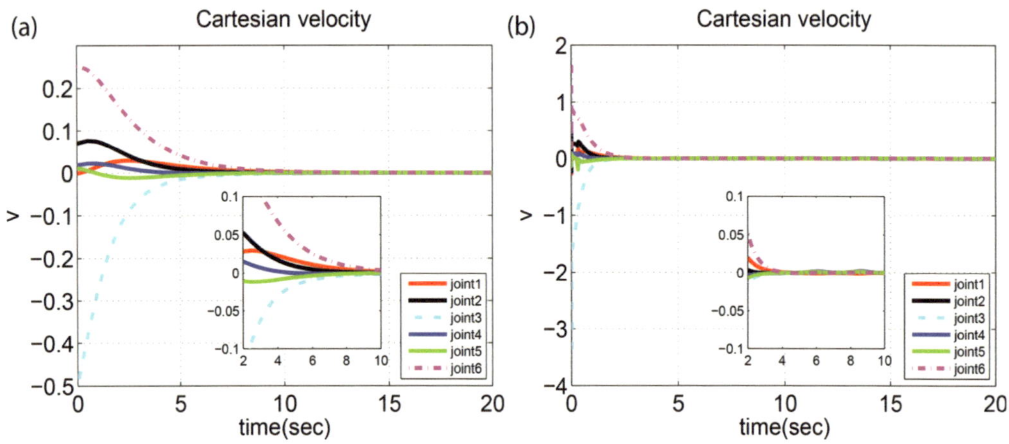

Figure 6. Joint velocity variation in pure rotation test. (a) IBVS and (b) enhanced IBVS (PD-SMC).

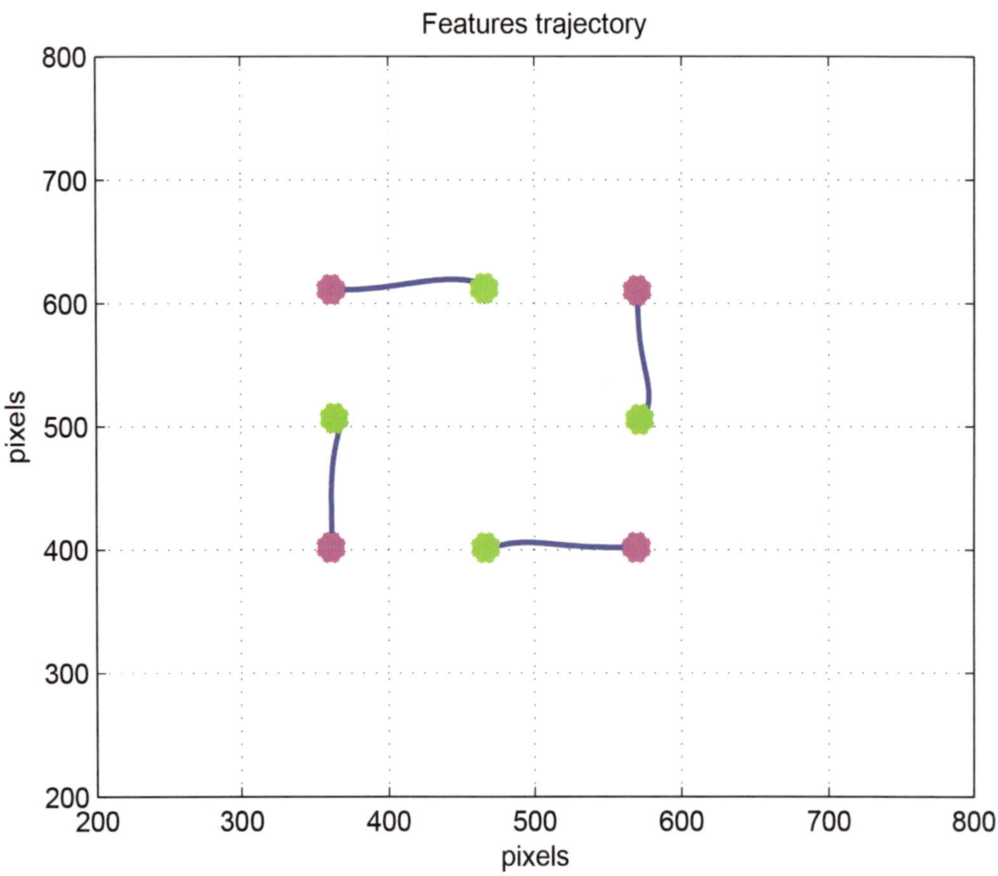

Figure 7. Feature trajectory in image space in Test 2.

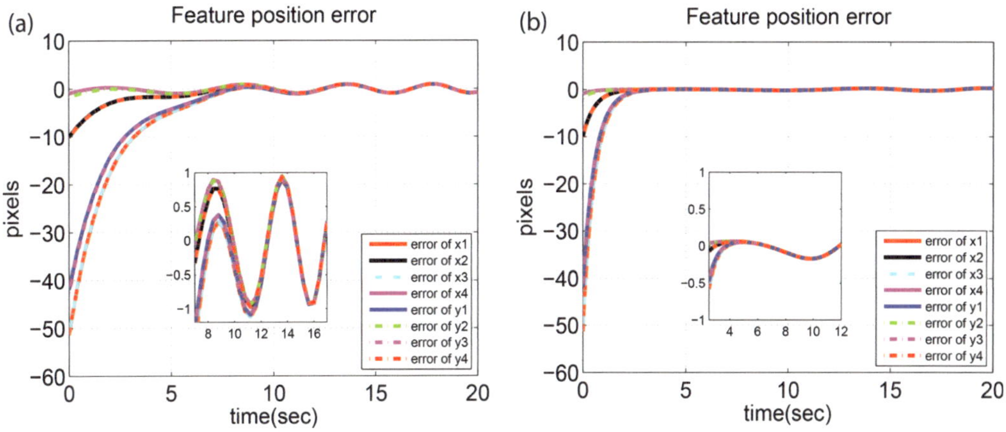

Figure 8. Feature error variation with disturbance. (a) IBVS and (b) enhanced IBVS (PD-SMC).

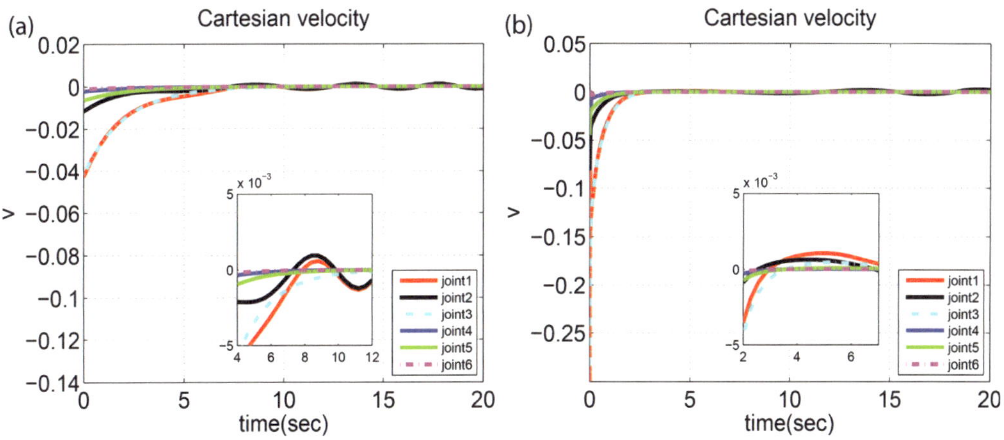

Figure 9. Joint velocity variation with disturbance. (a) IBVS and (b) enhanced IBVS (PD-SMC).

	IBVS	Enhanced IBVS
Test 1: ISE total	1.7875×10^4	5.3609×10^3
Test 2: ISE total	4.5601×10^5	1.6251×10^5
Test 3: ISE total	1.7639×10^4	5.3348×10^3

Table 3. ISE values of IBVS and enhanced IBVS.

6. Experiments

To further validate the performance of the proposed method, experimental tests are conducted on a 6DOF Denso robot. The experimental setup consists of a controller and two computers through a double PC bilateral teleoperation. PC No. 1 (Master PC) communicates with the master robot and transmits the commands to the Slave PC (PC No. 2) over the communication network. The slave PC also communicates with the slave robot (Denso robot) and obtains the camera data and sends it back to the master PC over the communication network in Refs. [17, 18]. The experimental setup is shown in **Figure 10**. The experimental system is shown in **Figure 11**. Denso VP6242G is used as the manipulator in [8]; the characteristics of the camera are given in **Table 4**.

Three experimental tests have been conducted, including long-distance translation and pure rotation of features, and hybrid translation-rotation test. Four feature points are used in visual servoing control. The initial and desired positions of the image features are given in **Table 5**. **Figure 12** shows Denso robot in initial position and in work processing.

Test 1 is performed to examine the convergence of image feature points when the desired position is far away from the initial one, which needs a long-distance translational motion. **Figure 13** shows that the feature position errors converge to zero. **Figure 14** shows the initial and desired positions captured by the camera. **Figure 15** shows the feature trajectory. **Figure 16** shows the camera trajectory in Cartesian space.

It is shown that the performance of hybrid PD-SMC is better than that of IBVS. The settling time of the hybrid PD-SMC method is shorter than that of conventional method. Furthermore, in hybrid PD-SMC method, the feature trajectory is straighter in image plane and the camera trajectory in Cartesian space is smoother.

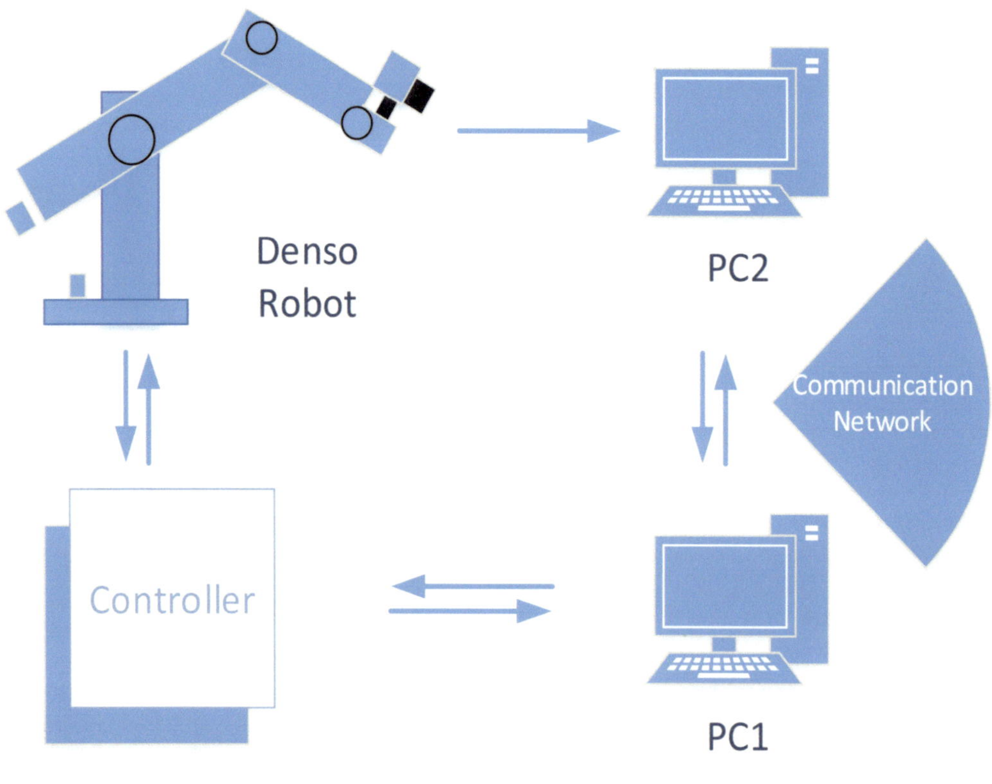

Figure 10. Experimental setup.

http://dx.doi.org/10.5772/67521

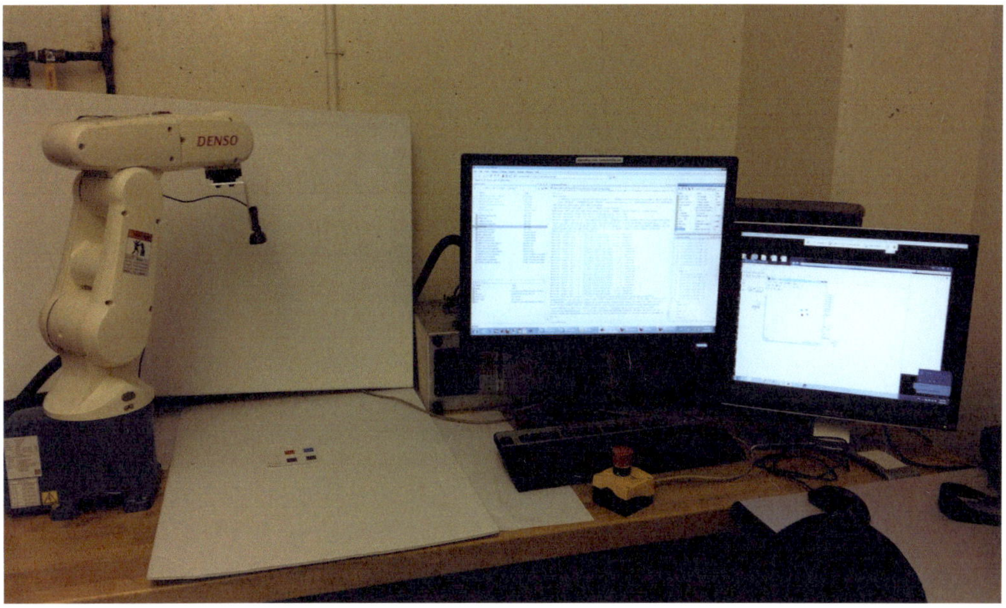

Figure 11. Experimental system.

Parameters	Values
Focal length	0.004 (m)
X-axis scaling factor	110,000 (pixel/m)
Y-axis scaling factor	110,000 (pixel/m)
Image plane offset of *X*-axis	120 (pixel)
Image plane offset of *Y*-axis	187 (pixel)

Table 4. Camera parameters in experiments.

	Positions			
	$(x_1\ y_1)$	$(x_2\ y_2)$	$(x_3\ y_3)$	$(x_4\ y_4)$
Test 1				
Initial	(57 150)	(57 57)	(146 63)	(146 148)
Desired	(595 270)	(595 175)	(684 177)	(686 275)
Test 2				
Initial	(454 213)	(385 146)	(447 81)	(516 148)
Desired	(602 270)	(600 174)	(688 179)	(619 273)
Test 3				
Initial	(103 136)	(196 105)	(225 187)	(134 220)
Desired	(447 203)	(540 189)	(557 278)	(461 292)

Table 5. Initial and desired positions.

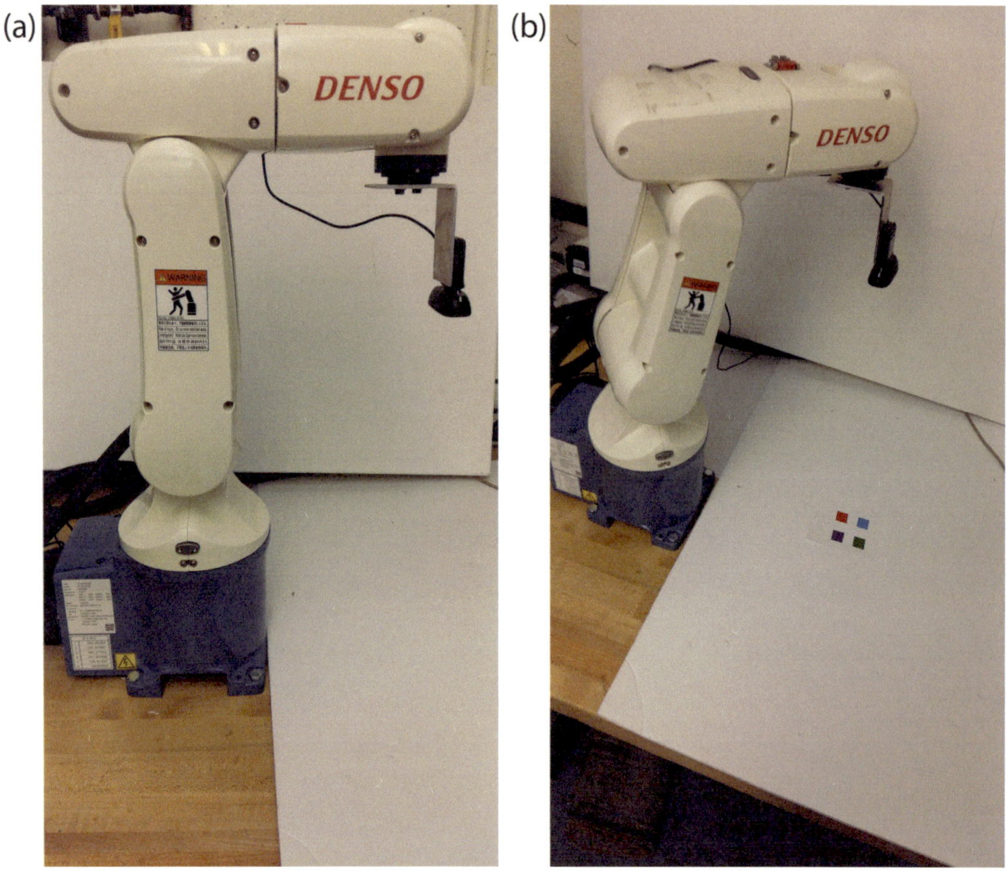

Figure 12. Denso robot. (a) Initial position and (b) work process.

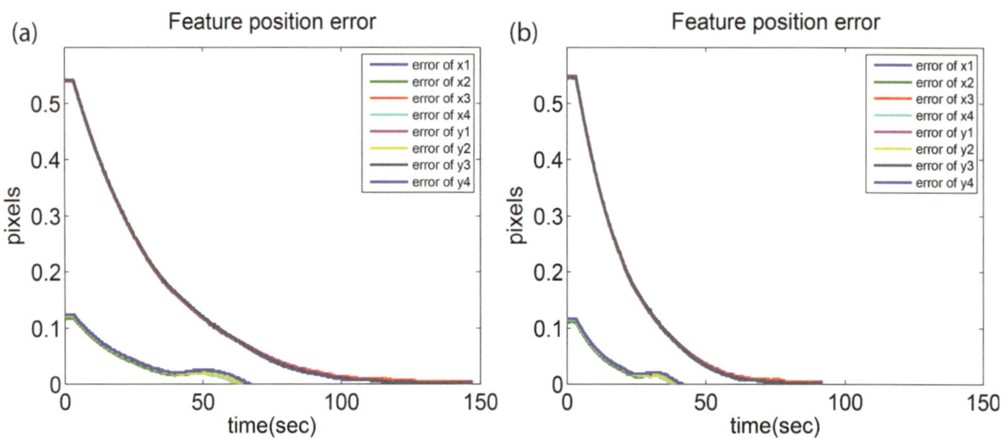

Figure 13. Feature error variation. (a) IBVS and (b) enhanced IBVS (PD-SMC) in Test 1.

http://dx.doi.org/10.5772/67521

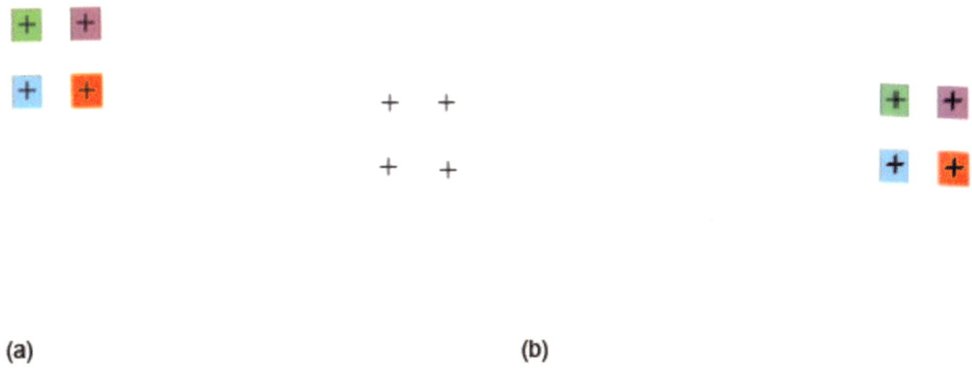

Figure 14. Feature position variation. (a) IBVS and (b) enhanced IBVS (PD-SMC) in Test 1.

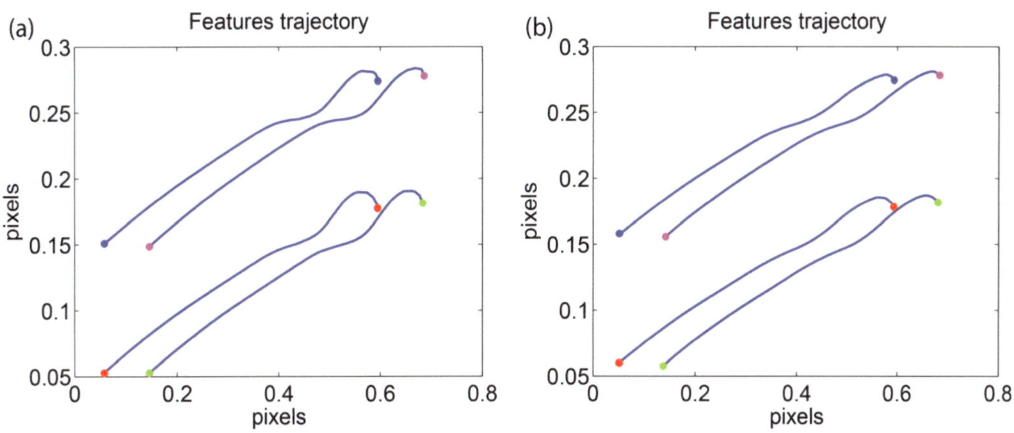

Figure 15. Feature trajectory. (a) IBVS and (b) enhanced IBVS (PD-SMC) in Test 1.

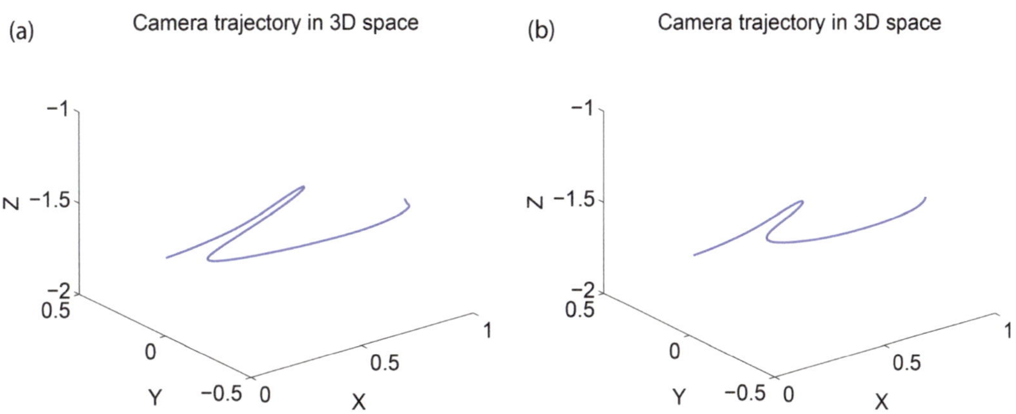

Figure 16. Camera trajectory in Cartesian. (a) IBVS and (b) enhanced IBVS (PD-SMC) in Test 1.

Test 2 is performed to examine the rotation performance of the proposed method; a pure rotation of image feature points has been completed. **Figure 17** shows that the feature position errors converge to zero. **Figure 18** shows the initial and desired positions, which are captured by the camera. **Figure 19** shows the feature trajectory in image plane. **Figure 20** shows the camera trajectory in Cartesian space.

It is obvious that the test is successfully performed to prove the better performance of enhanced IBVS. **Figures 17–20** show the comparison of experiment results. The results are similar to those of Test 1.

Test 3 is a hybrid translation-rotation motion process. In this experimental test, the translation and rotation motions of features are incorporated in one process. In the initial stage of the movement, the translation motion is implemented. In the final stage of the movement, the rotation motion is completed.

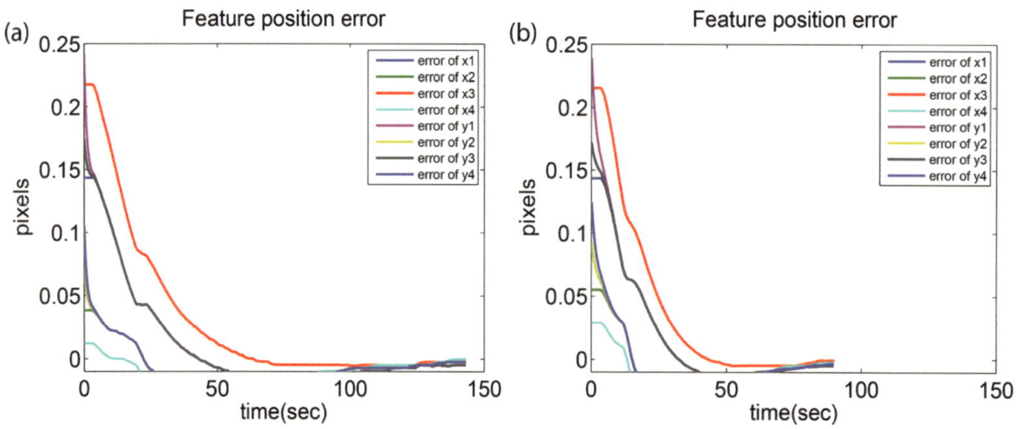

Figure 17. Feature error variation. (a) IBVS and (b) enhanced IBVS (PD-SMC) in Test 2.

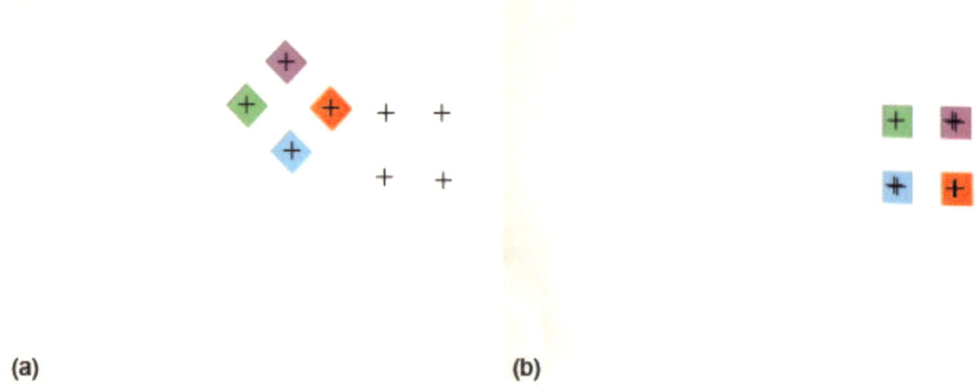

Figure 18. Feature position variation. (a) IBVS and (b) enhanced IBVS (PD-SMC) in Test 2.

http://dx.doi.org/10.5772/67521

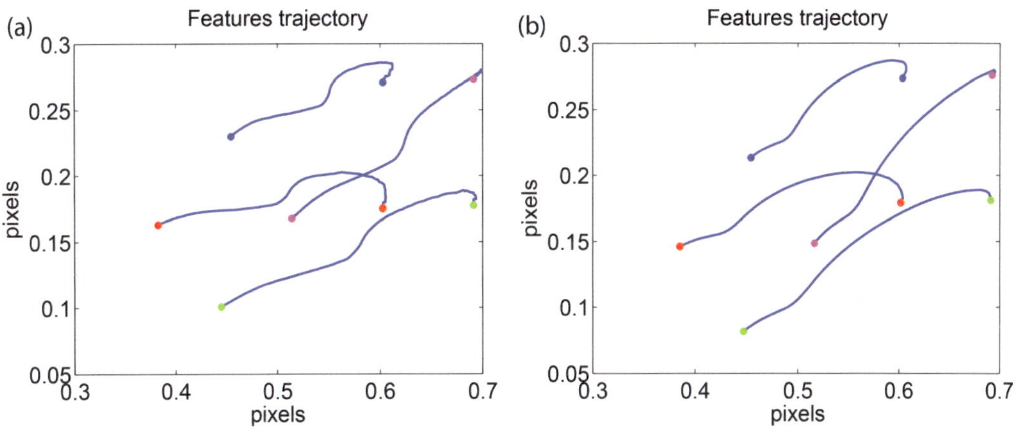

Figure 19. Feature trajectory. (a) IBVS and (b) enhanced IBVS (PD-SMC) in Test 2.

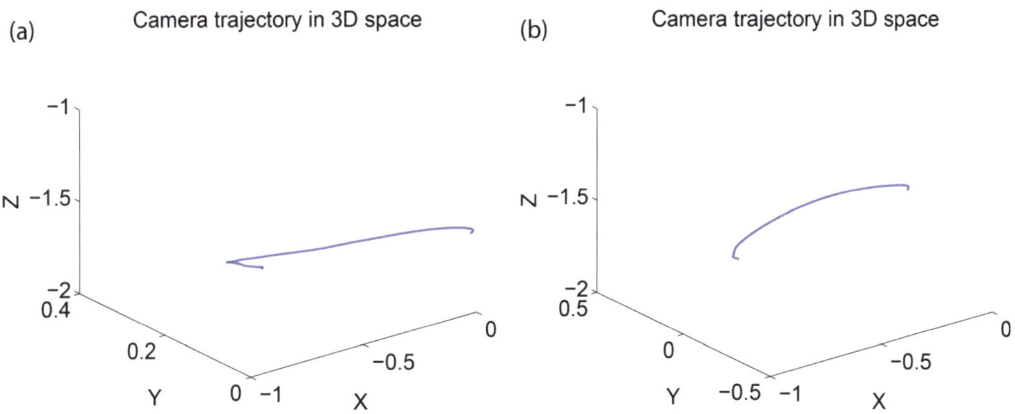

Figure 20. Camera trajectory in Cartesian. (a) IBVS and (b) enhanced IBVS (PD-SMC) in Test 2.

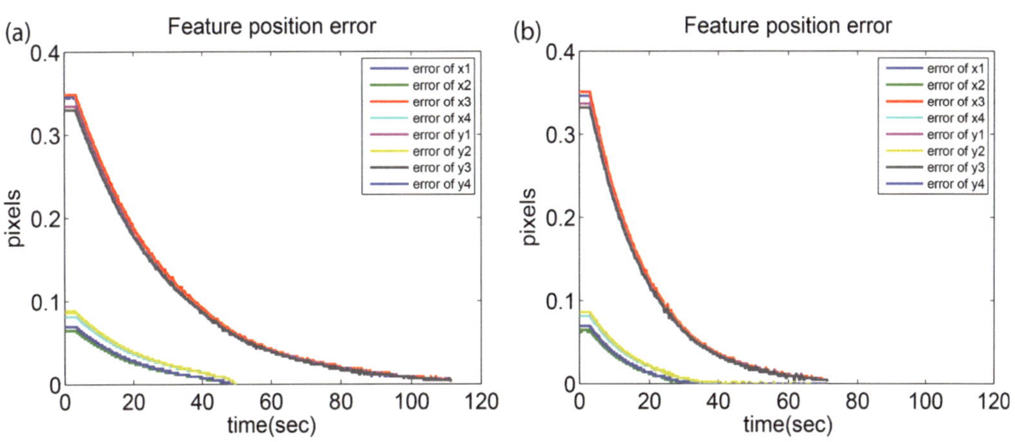

Figure 21. Feature error variation. (a) IBVS and (b) enhanced IBVS (PD-SMC) in Test 3.

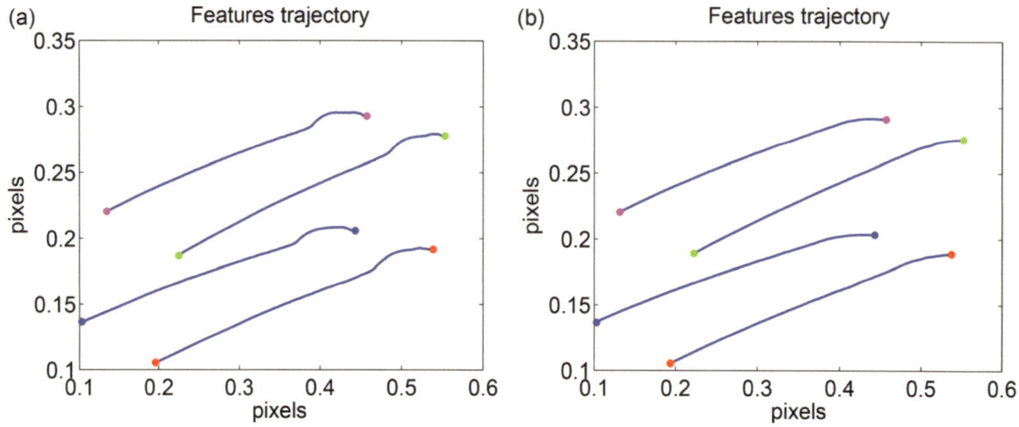

Figure 22. Feature trajectory. (a) IBVS and (b) enhanced IBVS (PD-SMC) in Test 3.

Figure 21 shows the feature position error variation of IBVS and enhanced IBVS. It is observed that enhanced IBVS owns the higher convergence rate. **Figure 22** shows the image feature points from the initial position to the final position and the trajectory by using IBVS and enhanced IBVS. It is observed that enhanced IBVS performs better in the final stage than IBVS in terms of the smoothness and length of its trajectories in image plane. **Figure 23** shows the camera trajectory in a three-dimensional space of IBVS and enhanced IBVS. It can be seen that the camera trajectory of enhanced IBVS is smoother and more accurate.

More specifically, the robustness against the random disturbances during the experiment is demonstrated in rotation movement. By comparing the trajectories, one can notice that the proposed enhanced IBVS method owns better robustness.

The performance index ISE (Integrate Square Error) is also used to compare the performance of IBVS and enhanced IBVS. The results are described in **Table 6**, and it shows that the ISE of enhanced IBVS is smaller than that of IBVS in three tests.

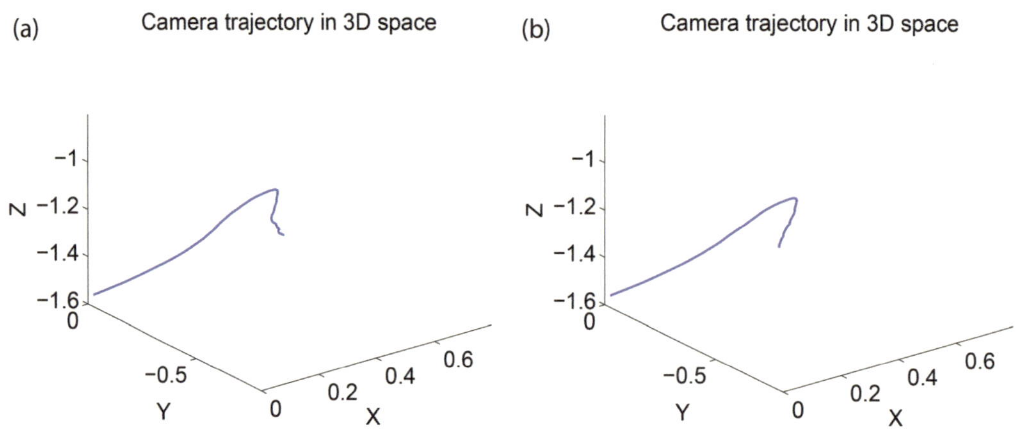

Figure 23. Camera trajectory in Cartesian. (a) IBVS and (b) enhanced IBVS (PD-SMC) in Test 3.

	IBVS	Enhanced IBVS
Test 1: ISE total	1160.0	791.5
Test 2: ISE total	112.6	92.7
Test 3: ISE total	438.6	296.8

Table 6. ISE values of IBVS and enhanced IBVS.

7. Conclusions

An enhanced IBVS, which combines PD control with SMC, is proposed for a 6DOF manipulator in this chapter. This approach can improve the visual servoing performance by taking the advantages of PD control and SMC and compensating for the shortcomings. The stability of the enhanced IBVS system is proven. Extensive simulations and experiments have been carried out and three tests are implemented for the comparison. The results validate that the tracking performance and robustness of the proposed method are superior to the conventional IBVS controller.

Author details

Shutong Li[1], Ahmad Ghasemi[2], Wen-Fang Xie[2]* and Yanbin Gao[1]

*Address all correspondence to: wfxie@encs.concordia.ca

1 College of Automation, Harbin Engineering University, Harbin, China

2 Department of Mechanical and Industrial Engineering, Concordia University, Montreal, Canada

References

[1] Ghanbari A, Wang W, Hann CE, Chase JG, Chen XQ. Cell image recognition and visual servo control for automated cell injection. In: 2009 4th International Conference on Autonomous Robots and Agents; 10–12 February 2009; IEEE; 2009. pp. 92–96.

[2] Hashimoto K. A review on vision-based control of robot manipulators. Advanced Robotics. 2003;**17**:969–991. DOI: 10.1163/156855303322554382

[3] Corke P. Robotics, Vision and Control. 1st ed. Springer Berlin Heidelberg; 2011. 495 p. DOI: 10.1007/978-3-642-20144-8

[4] Chaumette F, Hutchinson S. Visual servo control. I. Basic approaches. IEEE Robotics Automation Magazine. 2006;**13**:82–90. DOI: 10.1109/mra.2006.250573

[5] Hutchinson S, Hager GD, Corke P. A tutorial on visual servo control. IEEE Transactions on Robotics and Automation. 1996;**12**:651–670. DOI: 10.1109/70.538972

[6] Chaumette F, Hutchinson S. Visual servo control. II. Advanced approaches. IEEE Robotics Automation Magazine. 2007;**14**:109–118. DOI: 10.1109/mra.2007.339609

[7] Wang J, Cho H. Micropeg and hole alignment using image moments based visual servoing method. IEEE Robotics Automation Magazine. 2008;**55**:1286–1294. DOI: 10.1109/tie.2007.911206

[8] Keshmiri M, Xie WF, Mohebbi A. Augmented image-based visual servoing of a manipulator using acceleration command. IEEE Transactions on Industrial Electronics. 2014;**61**:5444–5452. DOI: 10.1109/tie.2014.2300048

[9] Allibert G, Courtial E, Chaumette F. Predictive control for constrained image-based visual servoing. IEEE Transactions on Robotics. 2010;**26**:933–939. DOI: 10.1109/tro.2010.2056590

[10] Kim JK, Kim DW, Choi SJ, Won SC. Image-based visual servoing using sliding mode control. 2006 SICE-ICASE International Joint Conference; 18–21 October 2006; IEEE; 2007. P. 4995–5001.

[11] Fang Y, Liu X, Zhang X. Adaptive active visual servoing of nonholonomic mobile robots. IEEE Transactions on Industrial Electronics. 2012;**59**:486–493. DOI: 10.1109/tie.2011.2143380

[12] Slotine JJE, Li WP. Applied Nonlinear Control. Prentice Hall; Englewood Cliffs, New Jersey 07632, 1991.

[13] Yksel T. IBVS with fuzzy sliding mode for robot manipulators. 2015 International Workshop on Recent Advances in Sliding Modes (RASM); 9–11 April 2015; IEEE; 2015. P. 1–6.

[14] Parsapour M, RayatDoost S, Taghirad HD. Position based sliding mode control for visual servoing system. 2013 First RSI/ISM International Conference on Robotics and Mechatronics (ICRoM); 13–15 February 2013; IEEE; 2013. P. 337–342.

[15] Acob JM, Pano V, Ouyang PR. Hybrid PD sliding mode control of a two degree-of-freedom parallel robotic manipulator. 2013 10th IEEE International Conference on Control and Automation (ICCA); 12–14 June 2013; IEEE; 2013. P. 1760–1765.

[16] Ngo QH, Nguyen NP, Chi NN, Tran TH, Hong KS. Fuzzy sliding mode control of container cranes. International Journal of Control, Automation, and Systems. 2015;**13**:419–425. DOI: 10.1049/iet-cta.2010.0764

[17] Corke P, Armstrong-Helouvry B. A meta-study of puma 560 dynamics: A critical appraisal of literature data. Robotica. 1995;13:253–258. DOI: 10.1017/s0263574700017781

[18] Spong MW, Hutchinson S. Robot Modeling and Control. Hoboken: Wiley; 2006. DOI: 10.1108/ir.2006.33.5.403.1

Chapter 4

Super Twisting Sliding Mode Control with Region Boundary Scheme for an Autonomous Underwater Vehicle

Vina Wahyuni Eka Putranti and Zool Hilmi Ismail

Additional information is available at the end of the chapter

http://dx.doi.org/10.5772/67579

Abstract

A robust tracking control for an Autonomous Underwater Vehicle (AUV) system operated in the extreme ocean environment activities is very much needed due to its external disturbances potentially disturb the stability of the system. This research proposes a new robust-region based controller which integrates Super Twisting Sliding Mode Control (STSMC) with region boundary approach in the presence of determined disturbances. STSMC is a second order SMC which combines between continuous signal and discontinuous signal to produce a robust system. By incorporating region based control into STSMC, the desired trajectory defined as a region produces an energy saving control compared to conventional point based control. Energy function of region error is applied on the AUV to maintain inside the desired region during tracking mission, thus, minimizing the energy usage. Analysis on a Lyapunov candidate proved that the proposed control achieved a global asymptotic stability and showed less chattering, providing 20s faster response time to handle perturbations, less transient of thrusters' propulsion and ability to save 50% of energy consumption compared to conventional SMC, Fuzzy SMC and STSMC. Overall, the newly developed controller contributed to a new robust, stable and energy saving controller for an AUV in the presence of external disturbances.

Keywords: super twisting sliding mode control, region boundary scheme, AUV

1. Introduction

An autonomous underwater vehicle (AUV) is employed with the aim to reduce the possibility of human accident in a long-term underwater mission. One of the important parts to be installed on an AUV is an advance control system. Beside the capability to ensure the robustness and efficiency of an AUV, the selected control system must have the capability to minimize the effect of hazardous underwater environment such as sea current and hydrodynamics forces that potentially increase the energy usage since the position of AUV is moved from the desired trajectory. Various controllers are introduced to be adapted on the underwater vehicle. Linear quadratic regulator (LQR), linear quadratic Gaussian (LQG), H_2, and H_∞ are examples of optimal control used to design a method which optimize some desirable parameters. In Ref. [1], Joshi and Talange compared PID and linear quadratic regulator (LQR) to control the depth of REMUS 100 AUV. It was shown that the PID control took faster time response compared to LQR but it produced greater overshoot. The steady-state error was not produced by both controllers. Wadoo et al. proposed an optimal feedback control, H_2, for trajectory tracking case of kinematic model on an AUV [2]. In this chapter, H_2 was obtained by formulating LQG as a system of two-norm optimization problem. For the result, the proposed control showed an optimal design since it proved the robustness to the disturbances. However, the dynamics model of the AUV was not included. The types of AUV as well as the types of disturbances were also not mentioned.

Meanwhile, other researchers employed a sliding mode controller (SMC) which is robust against an inaccurate model and the external disturbances [3]. In this case, Cristi et al. designed the SMC from the Lyapunov candidate then applied it to adjust the AUV's maneuver based on the dynamics system and operating condition [4]. The simulation obtained a small error but it did not include the effect of the disturbances. The integrator SMC was also proposed by Hong et al. for the depth control of torpedo-shaped AUV in Ref. [5]. The SMC was applied as an inner pitch controller, while the effect of buoyancy on pitch and heave dynamics was considered. It was shown that the steady-state error existed and bounded within 0.15 m. Akcakaya et al. simulated the SMC based on the Lyapunov candidate to observe the yaw steering of the NPS AUV II model [6]. The effect of the disturbances was added in the simulation, and it was assumed that the AUV moved along x-axis with a constant speed of 0.75 ms^{-1}. The AUV completed the task even if disturbances were introduced. However, the chattering effect was produced in the switching condition when the system tried to reach the sliding surface of the SMC. This caused overconsumption of energy and could damage the AUV because the rudder changed rapidly [6].

Fuzzy logic control (FLC) is well known as an intelligent and adaptive control method [7]. For some cases, FLC is used to solve the chattering problem, thus, Guo et al. superposed SMC with fuzzy tuning technique [8]. Stability and robustness of the control system were guaranteed by selecting the shrinking and dilating factors of the fuzzy membership functions. Two experiments were conducted to observe the efficiency of the proposed controller for a Hai-Min underwater vehicle under the influence of ocean current. The results confirmed the effectiveness of the proposed scheme, although a poor transient performance was produced when the system tried to achieve precise tracking. At the transition moment, the state and sliding surface were separated by a significant distance. Lakehekar and Saundarmal developed an adaptive fuzzy sliding mode controller with a boundary layer scheme [9]. In the simulation, the SMC was required to manage the vertical position of the

AUV. However, the variables inside SMC changed dramatically. Thus, a boundary layer near the switching line was introduced as a new method. To maintain the states inside the layer, two fuzzy approximators were employed. The first approximator was used to update the slope's value in the sliding surface, while the second approximator was used to shape the error tracking. The result showed that the proposed controller reduced the reaching time of 1–2 s faster in overcoming the perturbations compared to the conventional SMC and the fuzzy SMC. However, better results were obtained after formulating good parameter conditions, which were produced by creating many rules. Moreover, the use of many rules increased the energy demand.

Neural network (NN) is commonly used either as a control plant model or as a controller [7, 10]. There are two kinds of learning processes in the NN, online learning and offline training, and the success of NN depends on selecting the correct learning process. Some cases reported that different responses could be resulted even after the same controller was applied under the same environment [11]. Ji-Hong et al. compared the conventional SMC with neural network (NN) SMC [12]. When the SMC is widely used as a robust control, NN is used in conjunction to minimize the nonlinearity of the dynamic's error [13]. The results show that the NN produced small errors and the AUV was able to track the trajectory after many learning and adaptation processes [14]. The effect of the disturbances was also not considered. Meanwhile, Van de ven et al. approximated the damping model of an AUV by using the value of velocity and acceleration under offline training process [10]. Noise was added in the second simulation, hence an online learning was adopted to decrease the state prediction error as well as to minimize the influence of the noise. It was shown that NN was used to improve the performance of poor identification of the AUV model. However, Van de ven et al assumed that other parts of the AUV model to be fully known, while in real case, the other parts such as added mass and also colioris and centripetal model were fully uncertain.

Another robust controller, which has been developed has a high order sliding mode controller (HOSMC), works on higher order derivatives of the sliding variable/system deviation [15–19]. The development of this method aimed to minimize the chattering effect produced by the conventional SMC. The second order is widely implemented because of the low information demand. Robust integral of sign of error (RISE) is included in the type of HOSMC. Fischer et al. used the RISE as a robust control of a six-DOF AUV [20]. The experimental setup was conducted in a swimming pool and an openwater sea trial with 0.08 ms^{-1} of flow current. It was shown that the RISE gave a good performance despite larger orientation error being produced in an openwater sea trial. Then, Fischer et al. superposed the RISE with NN to solve the issue of dynamics model error by using online learning technique [21]. The simulation result showed that the error converged 10 s faster under the proposed control. Experimental validation was the next plan for a further research. Rhif proposed second-order sliding mode control, named 2-SMC, to control the position and speed of a torpedo AUV [22]. The presence of external disturbances was considered in the simulation, although its value and type were not mentioned in detail. Chattering effect was reduced by proposed control. In Ref. [23], 2-SMC was applied to observe the stability of cyclops AUV under constant value of disturbances. The proposed control reduced the effect of disturbances and the steady-state error. For further work, it was planned to develop the proposed control without decoupling some motions under sinusoidal disturbances.

Besides making a robust controller for the AUV, the other problem, which needs to be solved, was reducing the energy consumption. As stated in Ref. [6], the AUV could be damaged if it spends more energy. Li et al. was successful in introducing an adaptive region-tracking controller to overcome this weakness [24]. This success was followed by Ismail and Dunnigan [25]. The proposed controller in both the research guaranteed the error convergence of the sliding vector because the desired target was determined as a region instead of a point. The results showed that the thrusters were only activated when the AUV was outside the region. Therefore, the AUV reduced the energy consumption. Li and Cheah also used a similar approach for manipulator robot [26]. Li and Cheah proposed a unified objective bound method to merge the set point of the control, the trajectory tracking, and the performance bound. The desired trajectory reduced the conventional trajectory when the error was small, and it also changed to a dynamic region which could be scaled or rotated. Then, the system guaranteed the transient and the steady-state response of close loop system, as long as the objective was specified as a performance bound. A simulation was conducted to show the energy-saving properties of the proposed controller. The energy remained zero when the end effector of the arm robot started and stayed inside the bound. The proposed controller required less energy than the standard controller.

Reviewing from the advantages and the disadvantages of previous work, this chapter proposes a super twisting sliding mode control with region boundary for an AUV's tracking trajectory under the influence of perturbation. The proposed control is expected to obtain accuracy and efficiency, which is tracking precisely on the desired trajectory as well as saving energy consumption. The chapter is organized as follows: Section 2 studies about kinematic and dynamic model of a 6 DOF AUV, Section 3 describes the proposed control and comparison control, Section 4 performs results of simulation and analysis, while conclusion is explained in Section 5.

2. Kinematic and dynamic model

This section presents the kinematic and the dynamic model of a six-DOF AUV. Before discussing about the kinematic and the dynamic model, we introduce two types of geometric transformation in a six-DOF AUV, namely translation and rotations. The translation is represented by sway, surge, and heave movements, while the rotation is represented by roll, pitch, and yaw movements. An origin C, which is located on the center of the mass, a body-fixed reference and an earth-fixed reference, is used to describe the geometric transformation. The illustration can be seen in **Figure 1**.

The kinematic model studies about the relationship between inertial position of an AUV and velocity of an AUV. First, define the vector of position, vector of velocity, and vector of force as shown in Eq. (1) [27]:

http://dx.doi.org/10.5772/67579

$$\eta = [\eta_1; \eta_2]^T = [x, y, z; \phi, \theta, \psi]^T$$

$$v = [v_1; v_2]^T = [u, v, w; p, q, r]^T$$

$$\tau = [\tau_1; \tau_2]^T = [X, Y, Z; K, M, N]^T \tag{1}$$

where η indicates the linear and angular position, v indicates the linear and angular velocity, and τ indicates the linear and angular force. Jacobian matrix is used to approximate a small displacement in different spaces. Thus, the kinematic model from six DOF AUV is shown in Eq. (2) [27]:

$$\begin{bmatrix} v_1 \\ v_2 \end{bmatrix} = \begin{bmatrix} J_1^{-1}(\eta_2) & 0_{3\times3} \\ 0_{3\times3} & J_2^{-1}(\eta_2) \end{bmatrix} \begin{bmatrix} \dot{\eta}_1 \\ \dot{\eta}_2 \end{bmatrix} \Leftrightarrow v = J^{-1}(\eta)\dot{\eta} \tag{2}$$

where

$$J_1^{-1}(\eta_2) = \begin{bmatrix} \cos\psi\cos\theta & -\sin\psi\cos\phi + \cos\psi\sin\theta\sin\phi & \sin\psi\sin\phi + \cos\psi\cos\phi\sin\theta \\ \sin\psi\cos\theta & \cos\psi\cos\phi + \sin\phi\sin\theta\sin\psi & -\cos\psi\sin\phi + \sin\theta\sin\psi\cos\phi \\ -\sin\theta & \cos\theta\sin\phi & \cos\theta\cos\phi \end{bmatrix}$$

$$J_2^{-1}(\eta_2) = \begin{bmatrix} 1 & 0 & \sin\theta \\ 0 & \cos\phi & \cos\theta\sin\phi \\ 0 & -\sin\phi & \cos\theta\cos\phi \end{bmatrix}$$

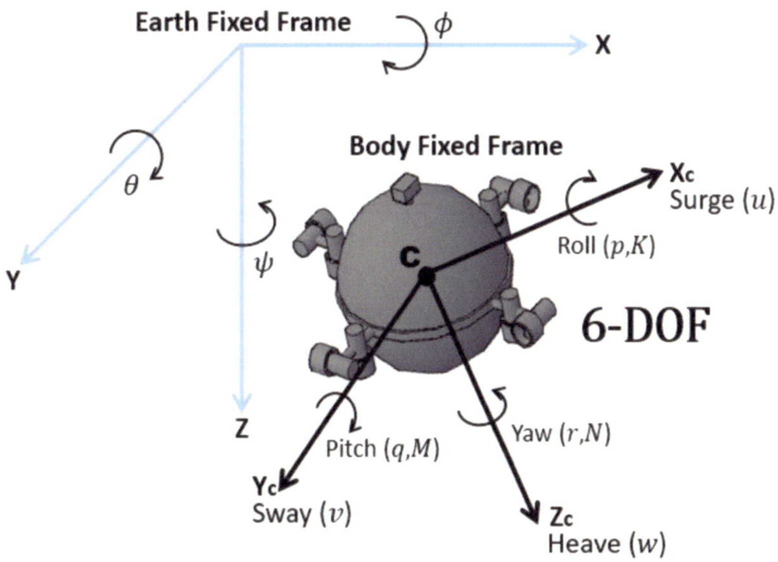

Figure 1. Body-fixed frame and earth fixed reference frame.

Meanwhile, the acceleration during a motion is studied by the dynamics model. The dynamic model is developed from the Newtonian and Lagrangian principles. This has been deeply discussed in Ref. [27]. The general dynamic model of the 6 DOFs AUV can be seen in Eq. (3):

$$M_{RB}\dot{v} + C_{RB}(v)v = \tau_{RB} \tag{3}$$

where $M_{RB} \in \mathbb{R}^{6\times 6}$ is the inertia matrix of a rigid body AUV, $C_{RB} \in \mathbb{R}^{6\times 6}$ is the coriolis and centripetal matrix of a rigid body, and $\tau_{RB} \in \mathbb{R}^{6\times 1}$ is an external force and moment. However, the ocean is a rough area. The hydrodynamics effects inside the ocean move and rotate the AUV from the initial position. The hydrodynamics effect is a force caused by fluids. These effects should be considered to avoid a model error of the AUV. The examples of hydrodynamics effects are radiation-induced forces and environmental forces. The equation of the forces of the hydrodynamics effect is established in Eq. (4) [27]:

$$-M_A\dot{v} - C_A(v)v - D(v)v - g(\eta) = \tau_H \tag{4}$$

where M_A is the added inertial matrix, C_A is the added hydrodynamics coriolis and centripetal matrix, D is the potential damping matrix, g is the gravitational matrix influenced by restoring forces, and τ_H donates the hydrodynamics' forces.

Eliminate the external forces by the hydrodynamics effect as shown in Eqs. (5–7):

$$\tau_{RB} - \tau_H = \tau \tag{5}$$

$$(M_{RB} + M_A)\dot{v} + (C_{RB} + C_A)(v)v + D(v)v + g(\eta) = \tau \tag{6}$$

$$M\dot{v} + C(v)v + D(v)v + g(\eta) = \tau \tag{7}$$

where $M \in \mathbb{R}^{6\times 6}$ indicates inertia matrix and added mass $(M_{RB} + M_A)$, $C(v) \in \mathbb{R}^{6\times 6}$ is the coriolis and centripetal matrix and added mass $\left(C_{RB}(v) + C_A(v)\right)$, $D(v) \in \mathbb{R}^{6\times 6}$ is the damping matrix (hydrodynamic damping and lift force), $g(\eta) \in \mathbb{R}^{6\times 1}$ represents the gravitational force and moment (restoring force), and $\tau \in \mathbb{R}^{6\times 1}$ is the control input/sum of estimated dynamics disturbances. Equation (2) is used to transform the dynamic model of AUV in Eq. (7) as follows,

$$M(\eta)\ddot{\eta} + C(v,\eta)\dot{\eta} + D(v,\eta)\dot{\eta} + g(\eta_2) = J^{-T}\tau \tag{8}$$

The dynamic model in Eq. (8) maintains Property 1, Property 2, and Property 3 [28].

Property 1 : M is symmetric and positive definite such that $M = M^T > 0$.

Property 2 : $C(v,\eta)$ is the skew-symmetric matrix such that $C(v,\eta) = -C^T(v,\eta)$.

Property 3 : $D(v,\eta)$ is positive definite, that is, $D(v,\eta) = D^T(v,\eta) > 0$.

3. Proposed control

This section discusses a proposed control which combines the super twisting sliding mode and region boundary scheme for a six-DOF AUV. Some equations in region boundary are used in super twisting sliding mode control; hence, this method is explained earlier.

3.1. Super twisting sliding mode controller

Super twisting is a part of the high order sliding mode control (HOSMC). The basic idea of HOSMC is removing the chattering effect, increasing the accuracy for tracking trajectory, and at the same time maintaining the advantages of conventional SMC [16]. In the conventional SMC, the control law consists of discontinuous system to ensure a sliding regime and the error convergence in a finite time happens when the system is restricted in the sliding surface. However, the high switching frequency known as chattering effect in the output signal is produced, thus, the stability of the control system is disturbed [29]. Furthermore, the value of sliding surface cannot be zero if the switching error exists. For this reason, super twisting SMC is used to preserve the zero value of sliding surface although in the presence of switching error [30].

There are two components in the super twisting SMC, the derivative of the discontinuous sliding surface and the continuous function of the sliding variable. Formulating the continuous function is useful to handle the chattering effect produced by the discontinuous function. The super twisting SMC is shown in Eq. (9):

$$\tau_{st} = \tau_1 + \tau_2 \tag{9}$$

where τ_1 denotes the discontinuous time derivative and τ_2 is a continuous function of the sliding variable. The values of τ_1 and τ_2 are determined in Eqs. (10) and (11), respectively:

$$\tau_1 = \int -\mathrm{K}\, sgn\,(s) \tag{10}$$

$$\tau_2 = -\kappa|s|^{0.5}\, sgn\,(s) \tag{11}$$

where s is the sliding surface, $\mathrm{K} \in \mathbb{R}$ represents a control parameter of the discontinuous system and its value is greater than zero, $\kappa \in \mathbb{R}$ represents a gain of continuous system, and sgn is a signum symbol. The value of $sgn\,(s)$ is equal to -1 if the sliding surface is less than zero, equal to zero if the sliding surface is zero, and equal to 1 if the sliding surface is greater than zero. The sliding surface is defined based on the first derivative of the tracking error, as shown in Eq. (12),

$$s = \dot{\eta} - \dot{\eta}_r \tag{12}$$

where $\dot{\eta}$ denotes the actual velocity and η_r denotes the reference vector which is developed from the region-based control. The final equation of super twisting sliding mode controller is given in Eq. (13):

$$\tau_{st} = \int -\mathrm{K}\, sgn\,(s) - \kappa|s|^{0.5}\, sgn\,(s) = \int -\mathrm{K}\, sgn\,(\dot{\eta} - \dot{\eta}_r) - \kappa|\dot{\eta} - \dot{\eta}_r|^{0.5}\, sgn\,(\dot{\eta} - \dot{\eta}_r) \tag{13}$$

Remark 3.1: Equation (13) reaches the finite time convergence as long as $\mathrm{K} > \frac{d}{\Gamma_M}$ and $\kappa^2 \geq \frac{4d\Gamma_M(\mathrm{K}+d)}{\Gamma_m^2\Gamma_m(\mathrm{K}+d)}$, where d is an arbitrary chosen as a positive real number of disturbance, Γ_m and Γ_M are constants with $\Gamma_m = \mathrm{K} - d_0$ and $\Gamma_M = \mathrm{K} + d_0$, while d_0 denotes the initial value of d.

3.2. Region boundary scheme

The region boundary scheme works by replacing line-based into region-based trajectory and different shapes of region can be decided by choosing the appropriate function [24]. The objective region in an inequality functions is given as follows,

$$f_i(\Delta\eta_i) = \begin{bmatrix} f_1(\Delta\eta_1) \\ f_2(\Delta\eta_2) \\ \cdots \\ f_N(\Delta\eta_N) \end{bmatrix} \leq 0 \tag{14}$$

where $\Delta\eta_i \triangleq (\eta - \eta_d) \in \mathbb{R}^3$, i is declared as 1, 2, …, N with N being the total number of objective function, $\eta \in \mathbb{R}^3$ is an actual position/orientation of an AUV, and $\eta_d \in \mathbb{R}^3$ is a reference point of $f_i(\Delta\eta_i)$. The actual position of an AUV is counted from the position of origin C of the vehicle. The example of region boundary scheme is explained as follows: the desired region is determined as a 2D square, the illustration is shown in **Figure 2**. The inequality function of **Figure 2** is given as

$$f_1(\Delta\eta_1) = (x - x_d)^2 - r_x^2 \leq 0$$
$$f_2(\Delta\eta_2) = (y - y_d)^2 - r_y^2 \leq 0 \tag{15}$$

where r_x and r_y are the regional bound.

To calculate the energy consumption when the AUV tracks on the region, the inequality function in Eq. (14) should be modified by adding the potential energy, given in Eq. (16):

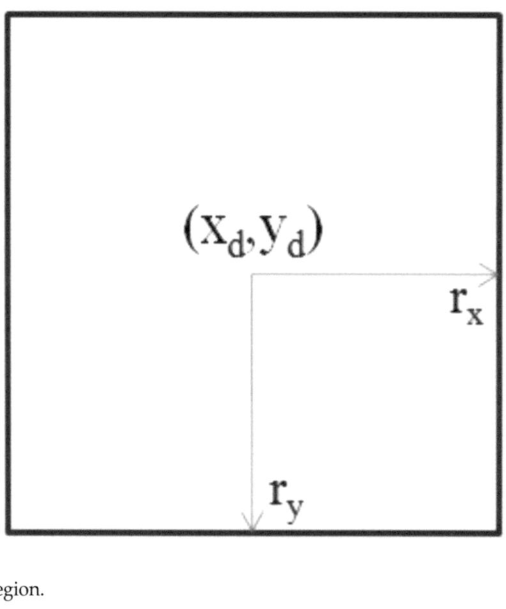

Figure 2. Rectangle desired region.

$$P(\Delta\eta) = \sum_{i=1}^{N} P_i(\Delta\eta_i) \tag{16}$$

where $P_i(\Delta\eta) \triangleq \frac{k_{pi}}{2}\left[\max(0,f_i(\Delta\eta))\right]^2$. The value of $P_i(\Delta\eta_i)$ relies on Eq. (17):

$$P_i(\Delta\eta_i) = \begin{cases} 0 & , \text{ if } f_i(\Delta\eta_i) \leq 0 \\ \frac{k_{pi}}{2} f_i^2(\Delta\eta) & , \text{ if } f_i(\Delta\eta_i) > 0 \end{cases} \tag{17}$$

where $k_p \in \mathbb{R}^{N\times N}$ denote positive constants of the potential energy. Note that the value of $f_i(\Delta\eta_i)$ is less than or equal to zero when the AUV enters the bound, thus, the gradient of $P_i(\Delta\eta_i)$ becomes smaller. Then, differentiating Eq. (17) with respect to $\Delta\eta_i$ yields Eq. (18):

$$\left(\frac{\partial P(\Delta\eta)}{\partial\Delta\eta}\right)^T = \sum_{i=1}^{N} k_{Pi}\max\left(0,f_i(\Delta\eta_i)\right)\left(\frac{\partial f_i(\Delta\eta_i)}{\partial\Delta\eta_i}\right)^T \triangleq \Delta e_\eta \tag{18}$$

where Δe_η is the region error whose value reduces to zero once the AUV move toward the desired region [24].

Remark 3.2: The region error will trigger an AUV toward the desired region. Once the AUV is inside the region, the gradient of potential energy, $P_i(\Delta\eta_i)$, becomes zero and at the same time Δe_η reduces smoothly to zero.

3.3. Super twisting sliding mode controller with region boundary scheme

Equation of super twisting SMC with region boundary scheme is shown in Eq. (19):

$$\tau = \tau_{st} + \tau_{eq} \tag{19}$$

where τ denotes a force acting on the center mass of an AUV or a control input, τ_{st} is a super twisting SMC, and τ_{eq} is the energy saving control law. Differentiating a sliding surface in Eq. (12) with respect to time yields Eq. (20):

$$\dot{s} = \ddot{\eta} - \ddot{\eta}_r \tag{20}$$

The following equation is the reference vector according to the region error,

$$\dot{\eta}_r = J^{-1}(\eta)(\dot{\eta}_d - \Delta\eta) - \alpha J^{-1}(\eta)\Delta e_\eta \tag{21}$$

where α is a constant value and $\Delta\eta$ represents the difference value between the actual and the desired position. Second derivatives of Eq. (21) with respect to time produces Eq. (22) [24]:

$$\ddot{\eta}_r = \dot{J}^{-1}(\eta)(\dot{\eta}_d - \Delta\eta) + J^{-1}(\eta)(\ddot{\eta}_d - \Delta\dot{\eta}) - \alpha\dot{J}^{-1}(\eta)\Delta e_\eta - \alpha J^{-1}(\Delta\dot{e}_\eta) \tag{22}$$

Then, multiplying M into both sides of Eq. (20) yields

$$M\dot{s} = M\ddot{\eta} - M\ddot{\eta}_r \tag{23}$$

where $M\ddot{\eta} = J^{-T}\tau_{eq} - (C\dot{\eta} + D\dot{\eta} + g)$. Determine $\dot{s} = 0$, hence [24]

$$M\dot{s} = J^{-T}\tau_{eq} - (M\ddot{\eta}_r + C\dot{\eta} + D\dot{\eta} + g)\tau_{eq} = J^T(M\ddot{\eta}_r + C\dot{\eta} + D\dot{\eta} + g) \tag{24}$$

Substitute $J^T \Delta e_\eta$ into Eq. (24), thus, energy saving potential control is obtained in Eq. (25) [24]:

$$\tau_{eq} = J^T(M\ddot{\eta}_r + C\dot{\eta} + D\dot{\eta} + g) - J^T \Delta e_\eta \tag{25}$$

Remark 3.3: Energy saving potential control drives the sliding surface converging to zero throughout the tracking mission, thus the AUV tracks inside the region.

Finally, super twisting SMC based on region boundary scheme is shown in Eq. (26):

$$\tau = \int \Big(-\mathrm{K}\, sgn\,(\dot{\eta} - \dot{\eta}_r) \Big) - \kappa |(\dot{\eta} - \dot{\eta}_r)|^{0.5}\, sgn\,(\dot{\eta} - \dot{\eta}_r) + J^T(M\ddot{\eta}_r + C\dot{\eta} + D\dot{\eta} + g) - J^T \Delta e_\eta \tag{26}$$

Theorem: The control input τ in Eq. (26) minimizes the chattering effect and allows the AUV to track on the desired region under determined perturbations as long as Remarks 3.1, 3.2, and 3.3 are fulfilled. Hence, the global asymptotic stability of closed loop systems is also guaranteed.

Proof: Propose a positive definite function of a Lyapunov candidate given in Eq. (27):

$$V = \frac{1}{2} s^T M s + \sum_{i=1}^{N} k_{P_i} \max\Big(0, f_i(\Delta\eta_i)\Big) \left(\frac{\partial f_i(\Delta\eta_i)}{\partial \Delta\eta_i}\right)^T + 2\kappa|s| + \frac{1}{2}\tau_1{}^2 + \frac{1}{2}\Big(\mathrm{K}|s|^{\frac{1}{2}}\, sgn\,(s) - \tau_1\Big)^2 \tag{27}$$

Define ζ equal to $\begin{bmatrix} |s|^{\frac{1}{2}}\, sgn\,(s) & \tau_1 \end{bmatrix}^T$ and $\overline{P}$ equal to $\frac{1}{2}\begin{bmatrix} 4\kappa + \mathrm{K}^2 & -\mathrm{K} \\ -\mathrm{K} & 2 \end{bmatrix}$. Equation (27) is transformed as a quadratic form as shown in Eq. (28):

$$V = \frac{1}{2} s^T M s + \sum_{i=1}^{N} k_{P_i} \max\Big(0, f_i(\Delta\eta_i)\Big)(\Delta\dot{\eta})^T \left(\frac{\partial f_i(\Delta\eta_i)}{\partial \Delta\eta_i}\right)^T + \zeta^T \overline{P} \zeta \tag{28}$$

Differentiating Eq. (28) with respect to time yields

$$\dot{V} = s^T M \dot{s} + \sum_{i=1}^{N} k_{P_i} \max\Big(0, f_i(\Delta\eta_i)\Big)(\Delta\dot{\eta})^T \left(\frac{\partial f_i(\Delta\eta_i)}{\partial \Delta\eta_i}\right)^T - \frac{1}{|s^{1/2}|}\zeta^T Q \zeta + q_1{}^T \zeta \tag{29}$$

where $Q = \frac{\mathrm{K}}{2}\begin{bmatrix} 2\kappa + \mathrm{K}^2 & -\mathrm{K} \\ -\mathrm{K} & 1 \end{bmatrix}$ and $q_1{}^T = \begin{bmatrix} 2\kappa + \frac{1}{2}\mathrm{K}^2 & -\frac{1}{2}\mathrm{K} \end{bmatrix}$. Substituting Eq. (24) into Eq. (29) gives [18]

http://dx.doi.org/10.5772/67579

$$\dot{V} = s^T\left(J^{-T}\tau - (M\ddot{\eta}_r + C\dot{\eta} + D\dot{\eta} + g)\right) + \sum_{i=1}^{N} k_{Pi}\max\left(0, f_i(\Delta\eta_i)\right)(\Delta\dot{\eta})^T\left(\frac{\partial f_i(\Delta\eta_i)}{\partial\Delta\eta_i}\right)^T - \frac{W}{2|s^{1/2}|}\zeta^T\tilde{Q}\zeta \tag{30}$$

where δ denotes coefficient of perturbation and $\tilde{Q} = \begin{bmatrix} 2\kappa + K^2 - \left(\frac{4\kappa}{K} + K\right)\delta & -K + 2\delta \\ -K + 2\delta & 1 \end{bmatrix}$.

Assume τ_{eq} in Eq. (25) as τ and substitute into Eq. (30), hence,

$$\dot{V} = -s^T(\Delta e_\eta) + \sum_{i=1}^{N} k_{Pi}\max\left(0, f_i(\Delta\eta_i)\right)(\Delta\dot{\eta})^T\left(\frac{\partial f_i(\Delta\eta_i)}{\partial\Delta\eta_i}\right)^T - \frac{K}{2|s^{1/2}|}\zeta^T\tilde{Q}\zeta \tag{31}$$

$$\dot{V} = -\left(\dot{\eta} - J^{-1}(\eta)(\dot{\eta} - \Delta\eta) - \alpha J^{-1}(\eta)\Delta e_\eta\right)(\Delta e_\eta) + \sum_{i=1}^{N} k_{Pi}\max\left(0, f_i(\Delta\eta_i)\right)(\Delta\dot{\eta})^T\left(\frac{\partial f_i(\Delta\eta_i)}{\partial\Delta\eta_i}\right)^T - \frac{K}{2|s^{1/2}|}\zeta^T\tilde{Q}\zeta \tag{32}$$

$$\dot{V} \leq -\alpha\Delta e_\eta{}^T\Delta e_\eta - \frac{K}{2|s^{1/2}|}\zeta^T\tilde{Q}\zeta \leq 0 \tag{33}$$

V is bounded since M is uniformly positively definite, hence, s and $P_i(\Delta\eta_i)$ are also bounded. By applying Barbalat's Lemma and Remark 3.1, it implies that $\dot{V}$ is negative definite if $\tilde{Q} > 0$. Therefore, the proposed control for the dynamic system of AUV in Eq. (7) guarantees $\Delta e_\eta \rightarrow 0$ and $s \rightarrow 0$ in $t \rightarrow \infty$. Region error converges to zero indicates that $f_i(\Delta\eta_i) \leq 0$, thus, $\frac{\partial f_i(\Delta\eta_i)}{\partial\Delta\eta_i}$ converges to zero.

3.4. Related control laws for comparative analysis

Because of the similarity of the proposed controller, some controller such as conventional sliding mode control (SMC), fuzzy SMC, and super twisting SMC are selected for the comparison purpose in the simulation. The equation of each comparison control is discussed as follows.

3.4.1. *Sliding mode control (SMC)*

The function of SMC is shown in Eq. (34) [5],

$$\tau_{SMC} = -K\, sgn\,(s) + \int k_s\, \dot{\eta}_r \tag{34}$$

The sliding surface s, $\dot{\eta}_r$, and $\Delta\eta$ are defined as follows

$$s = \dot{\eta} - \dot{\eta}_r \tag{35}$$

$$\dot{\eta}_r = J^{-1}(\eta)(\dot{\eta}_d - \Delta\eta) \tag{36}$$

$$\Delta\eta = \eta - \eta_d \tag{37}$$

where $k_s \in \mathbb{R}$ is a constant of integrate controller.

3.4.2. Fuzzy SMC

Formula of fuzzy SMC is given in Eq. (38) [31]

$$\tau_{\text{fuzzy}} = -K_f\, sgn\ (s) + \int k_s\, \dot{\eta}_r \tag{38}$$

where $K_f \in \mathbb{R}$ indicates the control parameter of discontinuous system which is obtained from the fuzzy rule, $k_s \in \mathbb{R}$ is a constant of an integrate controller, while the value of s, $\dot{\eta}_r$, and $\Delta\eta$ are the same as Eqs. (35), (36), and (37). The rule of fuzzy is given in **Table 1**, where s and $\dot{s}$ are the input of membership function and K_f is the output. Input s uses a trim type of membership function and its value varies from -150 to 150, input $\dot{s}$ uses the same type as s and its value varies from -1×10^{10} to 1.5×10^{10}, while output K_f uses *gauss2mf* as the type of membership function with range value from 5 to 25. The graph of the membership function is shown in **Figure 3**.

3.4.3. Super twisting SMC

The formula of super twisting SMC is given in Eq. (39) [32]

$$\tau_{\text{ST}} = \int -K\, sgn\ (s) - \kappa|s|^{0.5}\, sgn\ (s) \tag{39}$$

where $K \in \mathbb{R}$ is a control parameter of discontinuous system and $\kappa \in \mathbb{R}$ is a constant of continuous system. Equations (35–37) are used as the value of s, $\dot{\eta}_r$, and $\Delta\eta$.

		s						
K_f		**Negative large**	**Negative medium**	**Negative small**	**Zero**	**Positive small**	**Positive medium**	**Positive large**
$\dot{s}$	Negative	Large	Large	Large	Medium	Small	Medium	Large
	Zero	Large	Large	Medium	Small	Medium	Large	Large
	Positive	Large	Medium	Small	Medium	Large	Large	Large

Table 1. Fuzzy rule.

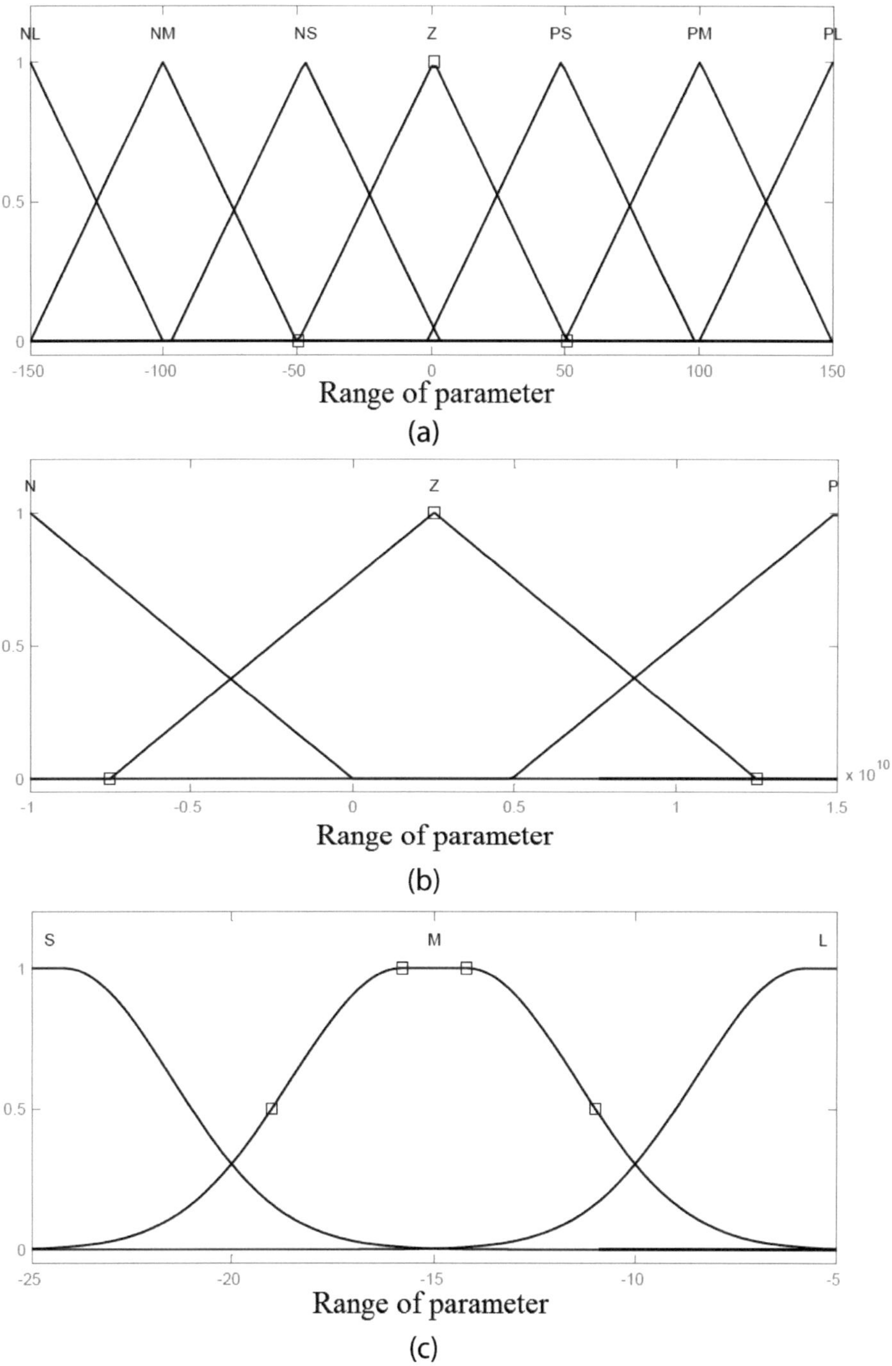

Figure 3. Membership function plot of fuzzy for (a) input s, (b) input $\dot{s}$ and (c) output K_f.

4. Results

The simulation utilizes an omni-directional intelligent navigator (ODIN) type of an AUV developed by Hawaii University as it is easy and widely used for the simulation purpose, as well as a holonomic model so that the singularity can be avoided and capable to move in six degree of freedoms (DOFs) without any preference of direction [33, 34]. The ODIN has eight thrusters to support its movement and does not require a heading angle to achieve a certain position. The horizontal diameter of an ODIN is 0.63 (m), while the vertical diameter is 0.61 (m). Its dry weight is about 125 (kg). There are two types of trajectories, which will be used in the simulation, a conventional line trajectory and a region trajectory. The conventional trajectory is applied on the conventional sliding mode control (SMC), fuzzy SMC, and super twisting SMC, while the region trajectory determined as a spherical shape is applied on the proposed control. An AUV is placed in the initial position at $[0 \quad 1 \quad 0]^{\mathrm{T}}$m, then it moves to the start sign at $[1.5 \quad 0 \quad -1.2]^{\mathrm{T}}$m, which indicates the start-tracking point, while the finish sign is at $[10 \quad 0 \quad 0]^{\mathrm{T}}$m where the simulation is stopped. The inequality equation of the spherical region is shown in Eq. (40) [25]:

$$f(\Delta\eta_1) = (x - x_d)^2 + (y - y_d)^2 + (z - z_d)^2 - r^2 \leq 0 \tag{40}$$

where $(x,\ y,\ z)$ represents the position of the vehicle in the x, y, and z axes, $(x_d,\ y_d,\ z_d)$ denotes the centre of the spherical region, and $r = 0.2$ m is the radius of the desired region. The value of the radius is determined arbitrary bigger than the radius of the ODIN, and there is no specific term on how to determine the value of the radius [24]. In the middle of the tracking activity, 0.05 ms^{-1} linear perturbation on x-axis, y-axis, and z-axis disturb the movement of an AUV. **Table 2** shows the parameter's values for each controller.

The first result is shown about tracking performance to see whether the AUV tracks outside the trajectory. As shown in the result of the conventional SMC in **Figure 4(a)**, the AUV reached the desired position after it moved down and gave oscillations of around 30 s at the start of the tracking point. Slightly similar to the conventional SMC, the AUV under the fuzzy SMC, shown in **Figure 4(b)**, also moved down before it reached the desired position, although the

No.	Controller	Parameter	
1.	Proposed controller	$\kappa = 14.5$	$K = 0.5$ $\alpha = 0.3$ $k_{pi} = \mathrm{diag}\{1,\ 1,\ 1\}$
2.	Conventional SMC	$K = 20$ $k_s = 0.01$	
3.	Fuzzy SMC	Input = s and $\dot{s}$ Output K_f from 5 to 25 $k_s = 0.01$	
4.	Super twisting SMC	$\kappa = 25$	$K = 5$ $k_s = 0.01$

Table 2. Technical description.

http://dx.doi.org/10.5772/67579

oscillations did not appear. A better performance was shown under the super twisting SMC which can be seen in **Figure 4(c)**. The AUV moved slightly straight toward the start tracking point from the initial position. The oscillations also did not appear under this controller. However, the proposed controller gave the best movement compared to the others. From **Figure 4(d)**, the AUV moved straight from the initial position to the start tracking point. In the perturbation time, it is shown that the higher the value of the perturbation, the further the AUV moved from its desired position. While **Figure 4(a)–(c)** show that the AUV was unable to maintain its position on the tracking line, the opposite result is shown in **Figure 4(d)**. The AUV remains inside the region boundary even though the perturbations are presented.

Next, results are discussed about error convergence and analyzed how long the controller takes the AUV to settle from the perturbation's effect. The time is counted from $\eta \neq 0$ to $\eta = 0$. $\eta \neq 0$ indicates the AUV is not on the desired position. In the case of the proposed controller, the error convergence is counted from $\eta \neq r$ to $\eta < r$. "r" sign indicates the radius of region boundary or the allowable error of the AUV's position based on Eq. (40), while $\eta < r$ indicates that the AUV has been inside the region. Another aim of error convergence is to see whether the controller produces the chattering effect during the convergence. **Figures 5(a)** and **(b)**

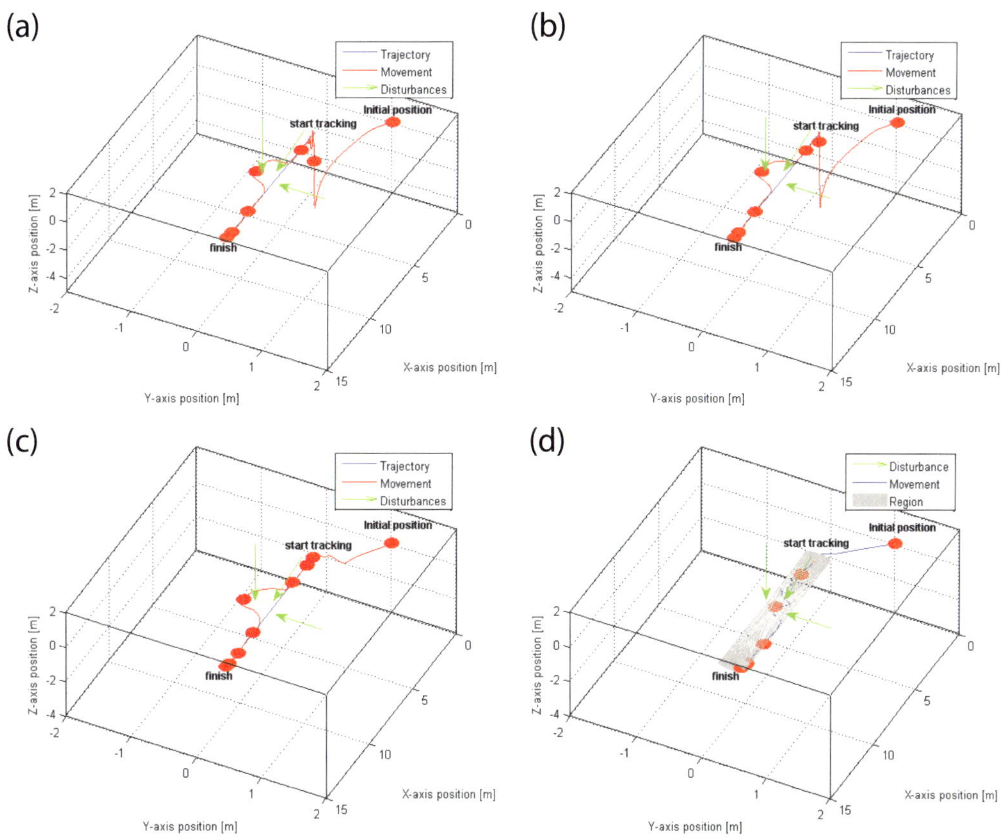

Figure 4. 3D results for (a) conventional SMC, (b) fuzzy SMC, (c) super Twisting SMC, and (d) proposed controller.

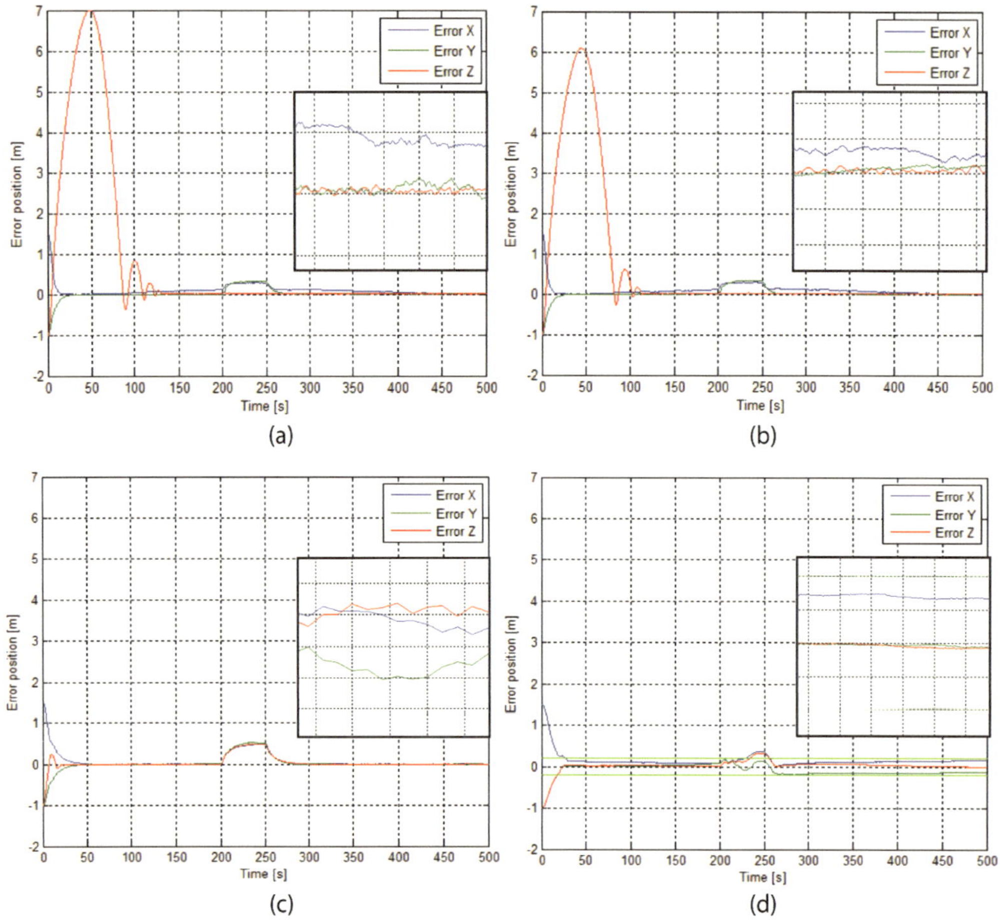

Figure 5. Error convergence for (a) conventional SMC, (b) fuzzy SMC, (c) super twisting SMC, and (d) proposed controller.

show that the X- and Y-axes of conventional SMC and fuzzy SMC converged to zero faster than the Z-axis in the beginning of the time. The X- and Y-axes of conventional SMC took 5 s and 125 s was needed by Z-axis. Meanwhile, the X- and Y-axes of fuzzy SMC required 27 s to make the AUV converge to the desired trajectory and 118 s were needed by the Z-axis. Conventional SMC and fuzzy SMC took a longer time to converge to zero due to the increasing value of perturbations. The time required for the error convergence in the X- and Y-axes was 180 s, while 15 s was required for the Z-axis. Meanwhile, the super twisting SMC required 30 s for the error convergence in all axes, while 5 s was required by the proposed controller during the perturbation time. The maximum error for the conventional SMC, fuzzy SMC, super twisting SMC, and the proposed controllers was 0.3, 0.3, 0.4, and 0.02 m, respectively. Meanwhile, the small graph of **Figures 5(a)** and **(b)** show the conventional SMC and fuzzy SMC producing the chattering in the presence of perturbation. A small chatter is also shown by super twisting SMC in small graph of **Figure 5(c)**, while proposed control showed a stable signal as performed in small graph of **Figure 5(d)**.

Results of force and moment are accumulated to obtain the energy spent by the AUV when the mission is executed. Force is an energy to make an object, in this case an AUV, move linearly from its initial position. On the other hand, moment is a force to make an AUV rotate from its center. The value of the force and moment were collected within the perturbed time or when the perturbations started to disturb the AUV. The reason for selecting this specific time was to see the differences in the way each controller handled the same perturbation. High amounts of force and moment indicate the inefficiency of the controller. The total amount was calculated by using the two-norm function or Euclidean distance and was considered as the energy spent by the AUV to accomplish the desired mission. As shown in **Table 3**, the proposed control spent the least energy at 356.72 N for force and at 65.02 Nm for moment compared to other controllers. The highest amount of force is spent by super twisting SMC at 463.45 N, while conventional SMC spent the highest amount of moment at 449.58 Nm.

The last results discuss about the thrusters' distribution. These data are collected to analyze the effort of the propeller to maintain the AUV at its desired position. The more effort spent by the propeller, the more power will be consumed. The ODIN had eight thrusters to move the AUV from one place to another. The eight thrusters were divided into two functions. Thrusters 1–4 were used to move the AUV horizontally, while thrusters 5–8 were used to move it vertically. **Figure 6** shows that the oscillation values of all the thrusters for conventional SMC and fuzzy SMC were similar. For super twisting SMC, more effort was required by thrusters 1, 3, 7, and 8 in the presence of disturbances, while the other thrusters showed constant oscillation. In the case of a proposed controller, all thrusters required more effort when the AUV was disturbed by perturbation.

It can be seen from the results that the proposed controller gave the best performance in terms of robustness and energy consumption. The first was referred from the value of the gain selection. The proposed controller needed smaller parameters value, 14.5 of κ and 0.5 of κ, compared to the other controllers. These parameters were used to trigger the movement of the AUV toward the desired position or were related to the tracking performance and error convergence. In normal situations, a high value of this parameter results in oscillations at the beginning of the time. In contrast, a small value of this parameter causes the AUV to take a longer time to reach the target. The worst case is the AUV never reaches the target. Second discussion is about error convergence. The results of the error convergence showed that all the controllers, except for the proposed controller, produced oscillations at the beginning of the time. The highest oscillation was produced by the conventional SMC, followed by the fuzzy SMC and the super twisting SMC. Hence, the AUV could not move directly from the initial position to the start tracking point under the conventional SMC and fuzzy SMC. Zero error convergences exist in the case of line trajectory for conventional SMC, fuzzy SMC, and super twisting SMC. Different result was shown in the use of region trajectory as the error position

Control	SMC	Fuzzy SMC	Super twisting SMC	Proposed control
Force	442.45 N	440.13 N	463.45 N	356.72 N
Moments	449.58 Nm	463.97 Nm	172.27 Nm	65.02 Nm

Table 3. Norm value of force and moment.

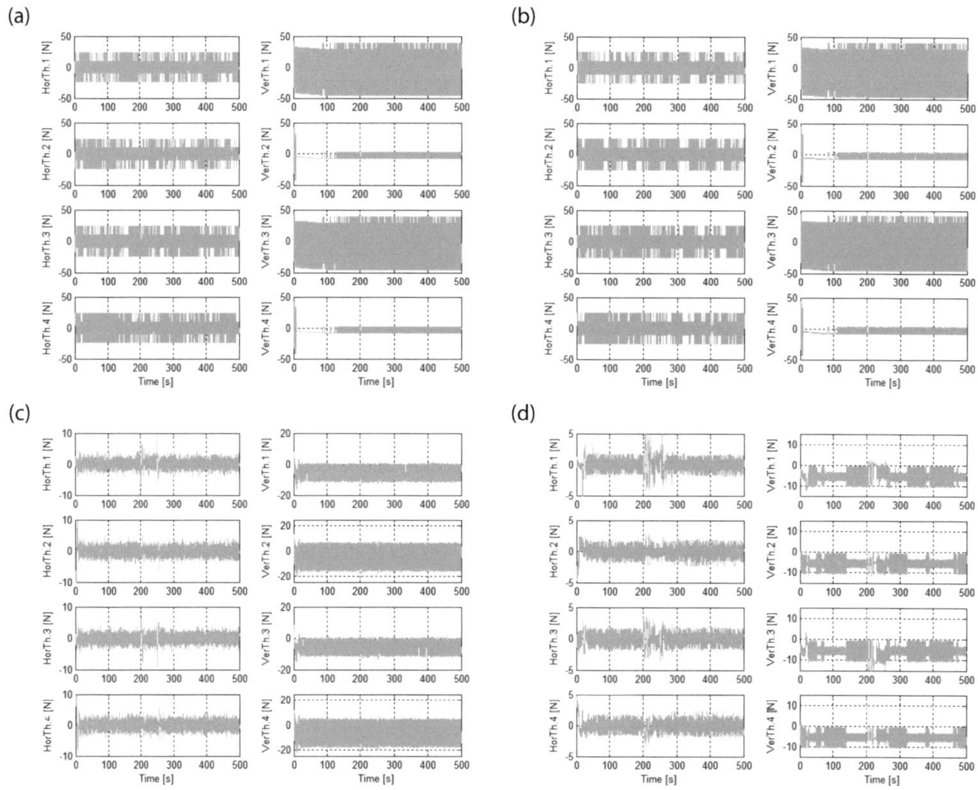

Figure 6. Thrusters' distribution for (a) conventional SMC, (b) fuzzy SMC, (c) super twisting SMC, and (d) proposed controller.

converged to the determined radius or allowable error instead of zero. This condition is in accordance with Remark 3.2 that as long as the AUV is inside the region, the gradient of potential energy becomes zero, thus the error region reduces to zero. According to the proof, one condition to ensure $s \rightarrow 0$ is the error convergence into region boundary.

In terms of the time required for the error convergence, the proposed controller took the shortest time to converge to zero, followed by the super twisting SMC. The position of the AUV at each axis under the conventional SMC and fuzzy SMC could not converge simultaneously. In this case, the time requirement stops had to be calculated when the position of the AUV in all the axes converged to zero. Thus, the conventional SMC took the longest time to converge to zero, followed by the fuzzy SMC. It could be seen from the results of the tracking performance that the AUV managed to move within the region under the proposed controller in the presence of perturbation. On the other hand, the other controllers failed to maintain the position of the AUV on the desired tracking line.

Based on the energy consumption, the super twisting SMC spent the highest force, both under constant and sine wave Gaussian white noise perturbations. The conventional SMC and fuzzy SMC were in the middle position, while the proposed controller spent the least force. The proposed control saved up to 30.2% of linear force consumption. For the moment under the constant perturbations, the highest value was spent by the conventional SMC, followed by the fuzzy SMC, the super twisting SMC, and the proposed controller. Meanwhile, the highest moment value under the sine wave Gaussian white noise perturbation was spent by the fuzzy SMC, followed by the conventional SMC, the super twisting SMC, and the proposed controller. In this case, the proposed controller saved more than 50% of the moment consumption.

More force was spent by the super twisting SMC and the proposed controller during the transition from the initial point. Therefore, the AUV was enabled to move directly to the start tracking point. These controllers also reacted to the presence of perturbations. The higher the value of the perturbation that disturbed the AUV, the greater the force that would be spent. In contrast, the conventional SMC and the fuzzy SMC showed a similar pattern of force and moment in all the times and in all conditions. This meant that these two controllers were not able to adapt well in handling changes in the situation.

The last discussion is about the propulsion or distribution of the thrusters. The proposed controller produced the least effort during the mission, followed by the super twisting SMC. Meanwhile, the conventional SMC and the fuzzy SMC showed the most active propulsion in all conditions. This situation was not good due to the battery consumption of the AUV. The more active thrusters indicated that the AUV would lose battery power easily. The high value of propulsion of the thrusters was also not good for the electrical devices inside the AUV.

5. Conclusion

A new robust-region-based controller is introduced from a survey of existing robust controls and saving energy approach for an autonomous underwater vehicle (AUV). Lyapunov candidate was used to prove a global asymptotical stability, while some simulations involving conventional sliding mode control (SMC), fuzzy SMC, and the only use of super twisting SMC were conducted under two kinds of perturbations to observe the effectiveness of the proposed controller. It is shown that the use of proposed controller was able to keep the AUV within the desired region under certain value of constant perturbations as well as a sinusoidal perturbation with a Gaussian white noise. From the results, it can be concluded that the proposed controller was able to minimize the chattering effect, provide a good response when overcoming the disturbances, provide a short computational time of error convergence, and save the amount of force and moment.

Author details

Vina Wahyuni Eka Putranti and Zool Hilmi Ismail*

*Address all correspondence to: zool@utm.my

Centre for Artificial Intelligence & Robotics, Universiti Teknologi Malaysia, Kuala Lumpur, Malaysia

References

[1] Joshi, S. D. and Talange, D. B. Performance Analysis: PID and LQR Controller for REMUS Autonomous Underwater Vehicle (AUV) Model. International Journal of Electrical Engineering & Technology (IJEET). 2012. 3(2): 320–327.

[2] Wadoo, S. A., Sapkota, S. and Chagachagere, K. Optimal Control of an Autonomous Underwater Vehicle. Systems, Applications and Technology Conference (LISAT), 2012. IEEE Long Island, Farmingdale, NY, 1–6.

[3] Nguyen, Q. H. and Kreuzer, E. A Robust Adaptive Sliding Mode Controller for Remotely Operated Vehicles. Technische Mechanik. 2007. 28(3–4): 185–193.

[4] Cristi, R., Papoulias, F. A. and Healey, A. J. Adaptive Sliding Mode Control of Autonomous Underwater Vehicles in The Dive Plane. IEEE Journal of Oceanic Engineering. 2002. 15(3): 152–160.

[5] Hong, E. Y., Soon, H. G. and Chitre, M. Depth Control of an Autonomous Underwater Vehicle, STARFISH. OCEANS 2010 IEEE. Sydney, 1–6.

[6] Akçakaya, H., Yildiz, H. A., Salam, G. and Gürleyen, F. Sliding Mode Control of Autonomous Underwater Vehicle. International Conference on Electrical and Electronics Engineering, 2009. ELECO 2009. November 5–8, 2009. Bursa: IEEE. 2009. II-332-II-336.

[7] Yildiz, O., Gokalp, R. B. and Yilmaz, A. E. A Review on Motion Control of the Underwater Vehicles. In: Proceedings of Electrical and Electronics Engineering, 2009. Bursa, 337–341.

[8] Guo, J., Chiu, F. C. and Huang, C. C. Design of a Sliding Mode Fuzzy Controller for The Guidance and Control of an Autonomous Underwater Vehicle. Ocean Engineering. 2003. 30(16): 2137–2155.

[9] Lakhekar, G. V. and Saundarmal, V. D. Novel Adaptive Fuzzy Sliding Mode Controller for Depth Control of an Underwater Vehicles. IEEE International Conference on Fuzzy Systems (FUZZ), 2013. July 7–10, 2013. Hyderabad: IEEE. 2013, 1–7.

[10] Van de Ven, P. W. J., Johansen, T. A., Sørensen, A. J., Flanagan, C. and Toal, D. Neural Network Augmented Identification of Underwater Vehicle Models. Control Engineering Practice. June 2007. 15(6): 715–725.

[11] Van de Ven, P. W. J., Flanagan, C. and Toal, D. Neural Network Control of Underwater Vehicles. Engineering Applications of Artificial Intelligence. August 2005. 18(5): 533–547.

[12] Li, J. H., Lee, P. M., Hong, S. W. and Lee, S. J. Stable Nonlinear Adaptive Controller for an Autonomous Underwater Vehicle Using Neural Networks. International Journal of Systems Science. 2007. 38(4): 327–337.

[13] Miao, B., Li, T. and Luo, W. A Novel Approach to Robust Adaptive NN Tracking Control for AUVs. 2014 33rd Chinese Control Conference (CCC), Nanjing. July 28–30, 2014. 8011–8016.

http://dx.doi.org/10.5772/67579

[14] Yuh, J. Learning Control for Underwater Robotic Vehicles. IEEE Control System Magazine. 1994. 14(2): 39–46.

[15] Levant, A. Higher-Order Sliding Modes, Differentiation and Output-Feedback Control. International Journal Control. 2003. 76(9–10): 924–941.

[16] Levant, A. Homogeinity Approach to High-Order Sliding Mode Design. Automatica. 2005. 41: 823–830.

[17] Moreno, J. A. and Osorio, M. A Lyapunov Approach to Second-Order Sliding Mode Controllers and Observers. 47th IEEE Conference on Decision and Control, 2008. CDC 2008. December 9–11, 2008. Cancun: IEEE. 2008. 2856–2861.

[18] Shtessel, Y. B., Moreno, J. A., Plestan, F., Fridman, L. M. and Poznyak, A. S. Super-Twisting Adaptive Sliding Mode Control: A Lyapunov Design. 49th IEEE Conference on Decision and Control, 2008. CDC 2008. December 15–17, 2010. Atlanta, GA: IEEE. 2010. 5109–5113.

[19] Rivera, J., Garcia, L., Mora, C., Raygoza, J. J. and Ortega, S. Super-Twisting Sliding Mode in Motion Control Systems. In: Bartoszewicz, A. (ed). Sliding Mode Control. Croatia: InTech. 239–254; 2011.

[20] Fischer, N., Hughes, D., Walters, P., Schwartz, E. M. and Dixon, W. E. Nonlinear RISE-Based Control of an Autonomous Underwater Vehicle. IEEE Transactions on Robotics. August 2014. 30(4): 845–852.

[21] Fischer, N., Bhasin, S. and Dixon, W. E. Nonlinear Control of an Autonomous Underwater Vehicle: A RISE-Based Approach. 2011 American Control Conference on O'Farrell Street, San Francisco, CA, USA. June 29–July 01 2011. 3972–3977.

[22] Rhif, A. A High Order Sliding Mode Control with PID Sliding Surface: Simulation on a Torpedo. International Journal of Information Technology, Control and Automation (IJITCA). January 2012. 2(1): 1–13.

[23] Khan, I., Bhatti, A. I., Khan, Q. and Ahmad, Q. Sliding Mode Control of Lateral Dynamics of an AUV. Proceedings of 2012 9th International Bhurban Conference on Applied Sciences & Technology (IBCAST). January 9–12, 2012. Islamabad: IEEE. 27–31.

[24] Li, X., Hou, S. P. and Cheah, C. C. Adaptive Region Tracking Control for Autonomous Underwater Vehicle. Control Automation Robotics & Vision (ICARCV), 2010 11th International Conference. December 7–10, 2010. Singapore: IEEE. 2129–2134.

[25] Ismail, Z. H. and Dunnigan, M. W. Tracking Control Scheme for an Underwater Vehicle-Manipulator System with Single and Multiple Sub-Regions and Sub-Task Objectives. Control Theory & Applications, IET. 2011. 5(5): 721–735.

[26] Li, X. and Cheah, C. C. Adaptive Neural Network Control of Robot Based on a Unified Objective Bound. IEEE Transactions on Control Systems Technology. 2014. 22(3). 1032–1043.

[27] Fossen, T. I. Handbook of Marine Craft Hydrodynamics and Motion Control. First Edition, New York: John Wiley and Sons. 2011.

[28] Ismail, Z. H., Faudzi, A. A. and Dunnigan, M. W. Fault-Tolerant Region-Based Control of an Underwater Vehicle with Kinematically Redundant Thrusters. Mathematical Problems in Engineering. 2014(2014): 1–12. http://dx.doi.org/10.1155/2014/527315

[29] Fridman, L. Chattering Analysis in Sliding Mode Systems with Inertial Sensors. International Journal of Control. 2003. 6(9-10). 906–912.

[30] Fridman, L. and Levant, A. Higher Order Sliding Modes. In: Perruquetti, W. and Barbot, J.-P. (eds). Sliding Mode Control in Engineering. USA: Marcel Dekker, Inc. 53–96; 2002.

[31] Putranti, V., Mokhar, B. and Ismail, Z. H. Sliding Mode Fuzzy Controller for Autonomous Underwater Vehicle under Deterministic Disturbances. Journal of Signal Processing. 2015. 19(4): 143–146.

[32] Salgado-Jimenez, T. and Jouvencel, B. Using a High Order Sliding Modes for Diving Control a Torpedo Autonomous Underwater Vehicle. OCEANS 2003 Proceedings. September 22–26, 2003. San Diego, CA, USA: IEEE. 934–939.

[33] Antonelli, G., Chiaverini, S., Sarkar, N. and West, M. Adaptive Control of an Autonomous Underwater Vehicle: Experimental Results on ODIN. IEEE Transactions on Control Systems Technology. 2001. 9(5): 756–765.

[34] Hong, E. Y., Soon, H. G. and Chitre, M. Depth Control of an Autonomous Underwater Vehicle, STARFISH. OCEANS 2010 IEEE. Sydney, 1–6.

Chapter 5

Adaptive Integral High-Order Sliding Mode for a Fixed Wing Aircraft

Zaouche Mohammed, Foughali Khaled and
Amini Mohamed

Additional information is available at the end of the chapter

http://dx.doi.org/10.5772/67580

Abstract

In order to develop and implement the laws piloting for an aircraft, flights validation will be necessary. This could in fact be done, in a first step, by using flight simulators. In this work, we choose the predator virtual model flying in Microsoft^{TM} flight simulator (MSFS) and we propose the procedure of controlling its attitude. We send the adaptive integral high-order sliding mode (AIHOSM) inputs piloting control. This work is a real-time virtual simulation. For the AIHOSM controller, we propose the gain adaptation for reduction of chattering phenomena and possibility to control the aircraft presented by the uncertain nonlinear systems in which the uncertainties have unknown bounds. This technique is more robust and simpler to implement than the quaternion one and only needs the information about the sliding mode surface.

Keywords: adaptive integral high-order sliding mode controller, Microsoft flight simulatorr, UAV predator, real-time virtual simulation

1. Introduction

In reality, all physical systems are affected by uncertainties due to modeling errors, parametric variation, and external disturbances. Controlling of dynamical systems in the presence of uncertainties is extremely difficult as the controller's performances degrade and the system may even be led to instability. As such, active researches are continuing to develop controllers that can work successfully in spite of uncertainties. Robust control techniques such as nonlinear adaptive control, model predictive control, backstepping and sliding mode control [1, 2, 3, 4, 5, 11, 19, 20, 32, 34] have been evolved to deal with uncertainties.

The classical Sliding Mode Control (SMC) leads, generally, to the appearing of an undesirable chattering phenomenon [2, 3, 9, 10, 13, 14, 15] to solve this problem we propose an approach using the Adaptive Integral High Order Sliding Mode Controller (AIHOSMC). This technique ensures a good tradeoff between error and robustness against noise and especially a good accuracy for a certain frequency range, regardless of the gain setting of the algorithm. This technique is based on estimating the successive derivatives of the sliding mode surface and transmitting them to the control block, all by using an aircraft in virtual simulated environments [24, 25]. It is real-time virtual simulation, which is close to the real-world situation.

The piloting technique proposed in this work is more robust and simpler to implement than the quaternion one. It only requires information about the sliding mode surface.

2. Problem statement

Through a methodology based on the confrontation of the real and the simulated worlds, the main objective of this chapter is to develop an autopilot based on a robust controller to maintain the desired trajectory (**Figure 1**).

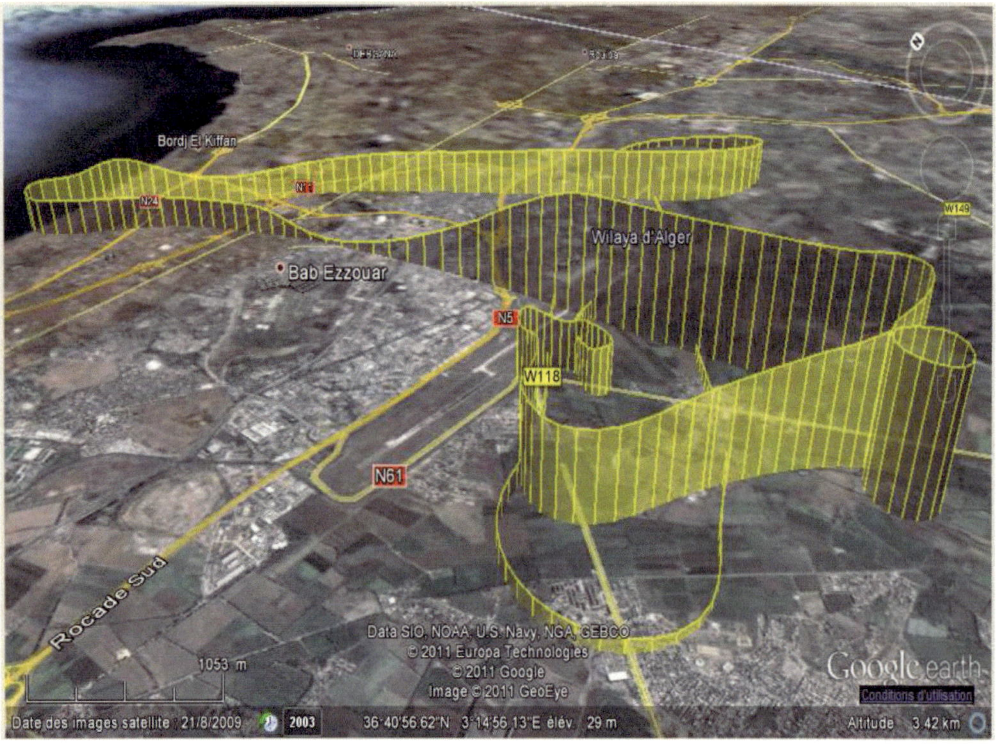

Figure 1. Real trajectory.

http://dx.doi.org/10.5772/67580

To achieve this objective, we use the flight simulator FS2004 as a simulated world environment coupled to a hardware and a software development platform. This simulator is developed by Microsoft, with several simulated aircraft included in its airplane library. We choose the Predator MQ-1 (**Figure 2**). It is considered as a reconnaissance and an intelligent system.

In this work, the main goal is to maintain the desired aircraft's trajectory; and to do so, we propose the following approach:

- description and analysis of the aircraft system model;
- implementation of a real-time interface between the flight simulator FS2004 and the module real-time Windows target of Simulink/Matlab;
- development and implementation of the piloting law based on adaptive integral sliding mode for the design of the autopilot controller;
- flight tests.

Figure 2. The predator MQ-1 flying in FS-2004.

3. Characteristics of the predator

The MQ-1 predator is an American unmanned aerial vehicle (UAV) that can serve in the reconnaissance or attack role. Predator has been in the United States Air Force (USAF) service since 1995 and has seen combat in numerous theatres.

Airwrench tool gives access to flight dynamic characteristics (http://www.mudpond.org/AirWrench_main.htm). This tool allows creating and tuning flight dynamics files description of simulated planes models. This software uses aerodynamics formulas and equations described on the Mudpond Flight Dynamics Workbook. It calculates aerodynamic coefficients based on the physical characteristics and performance of the aircraft (**Table 1**).

Dimensions	Moments of inertia
Length: 11.88 m	Pitch: 1800.0
Wingspan: 14.84 m	Roll: 3700.00
Wing surface area: 11.43 m^2	Yaw: 1800.00
Wing root chord: 1.55 m	Cross: 0.00
Aspect ratio: 19.28	
Taper ratio: 0.10	

Table 1. FS2004 aircraft-simulated characteristics PREADAR MQ-1.

4. Implementation of a real-time interface between Microsoft flight simulator and the module "real-time windows target" of Simulink/Matlab

We communicate with FS2004 by using a dynamic link library called FSUIPC.dll (Flight Simulator Universal Inter-Process Communication). This library created by Peter Dowson and is downloadable from his website [36] (*www.schiratti.com/dowson.html*). It allows external applications to read and write in and from Microsoft flight simulator (MSFS) by the means of an IPC (interprocess communication) using a buffer of 64 Ko. The documentation given with FSUIPC explains the organization of this buffer [8, 17, 18].

To read or write a variable using the FSUIPC, we need to know its offset address, its format, and the necessary conversions. For example, the bank angle (φ) is read as a signed long S32 at the offset 0x057C. **Table 2** shows the parameters used in our simulation.

To deal with the design of an autopilot controller, we propose an environment framework based on a software in the loop (SIL) methodology (see **Figure 3**) and we use Microsoft flight simulator (MSFS-2004) as a plane simulation environment [24, 25].

This work is a real-time virtual simulation, we read or/and write the desired parameters from and to MSFS-2004 through the computer memory by using the FSUIPC library.

Offset	Name	Var. type	Size (octet)	Usage
057C	Bank angle (φ)	S32	4	Degree
578	Elevation angle (θ)	S32	4	Degree
580	Head angle (ψ)	U32	4	Degree
02BC	Speed IAS (V)	S32	4	Knot*128
0BB2	Elevator deflection (δ_e)	S16	2	-16383 to +16383
0BB6	Aileron deflection (δ_a)	S16	2	-16383 to +16383
0BBA	Rudder deflection (δr)	S16	2	-16383 to +16383
088C	Thrust control (δ_x)	S16	2	-16383 to +16383

Table 2. Flight parameters in the buffer FSUIPC.

http://dx.doi.org/10.5772/67580

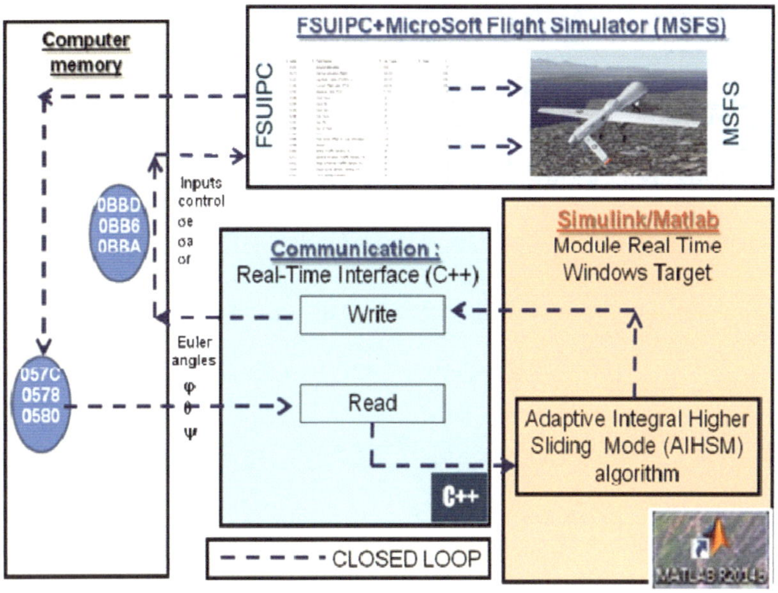

Figure 3. Software-in-the-loop architecture.

5. System modeling

The model describing the system is presented by [12, 25, 26]

$$\dot{x} = f(x) + g(x).U \tag{1}$$

with is the aircraft state vector in the body frame:

$$\begin{aligned} x &= [u \quad v \quad w \quad p \quad q \quad r \quad \varphi \quad \theta \quad \psi]^T \\ &= [x_1 \quad . \quad . \quad . \quad . \quad . \quad . \quad . \quad x_9]^T \end{aligned} \tag{2}$$

$U = [\delta_t \quad \delta_e \quad \delta_a \quad \delta_r]^T$ is the control vector and δ_t, δ_e, δ_a and δ_r denoting thrust control, elevator deflection, aileron deflection, and rudder deflection, respectively.

We propose the following output vector:

$$y = [\phi \quad \theta \quad \psi]^T \tag{3}$$

The nonlinear functions $f(x)$ and $g(x)$ are given by [16, 23, 25]:

$$f(x) = [f_1(x) \quad . \quad . \quad f_9(x)]^T \tag{4}$$

where,

$$f_1(x) = x_2x_6 - x_3x_5 + C_{x2}x_5 + C_{x4} + C_{x5}\alpha + C_{x1}\dot{\alpha} - g\sin x_8$$

$$f_2(x) = x_3x_5 - x_2x_4 + C_{y2}x_4 + C_{y3}x_6 + C_{y6}\beta + C_{y1}\dot{\beta} + C_{y7} + g\sin x_9 \sin x_8$$

$$f_3(x) = x_1x_5 - x_1x_2 + C_{z2}x_5 + C_{z5}\alpha + C_{z1}\dot{\alpha} + C_{z4} + g\cos x_9 \cos x_8$$

$$\begin{aligned} f_4(x) = &-\frac{I_{zz}}{\Delta}[-I_{xz}x_4x_5 + (I_{yy} - I_{zz})x_6x_5 + C_{l2}x_4 + C_{l3}x_6] \\ &-\frac{I_{xz}}{\Delta}[-I_{xz}x_6x_5 + (I_{yy} - I_{xx})x_4x_5 - C_{n2}x_4 + C_{n3}x_6] \\ &-\frac{1}{\Delta}[I_{zz}(C_{l5}\beta + C_{l1}\dot{\beta} + C_{l7}) - I_{xz}(C_{n6}\beta + C_{n1}\dot{\beta}) - C_{n7}] \end{aligned}$$

$$f_5(x) = -\frac{1}{I_{yy}}(I_{zz} - I_{xx})x_4x_6 + I_{xz}(x_6^2 - x_4^2) + C_{m2}x_5 + C_{m5}\alpha + C_{m1}\dot{\alpha} + C_{m4}$$

$$\begin{aligned} f_6(x) = &-\frac{I_{xz}}{\Delta}[I_{xz}x_4x_5 - I_{xz}(I_{yy} - I_{zz})x_6x_5 + C_{l2}x_4 + C_{l3}x_6] \\ &-\frac{I_{xx}}{\Delta}[-I_{xz}x_6x_5 + (I_{yy} - I_{xx})x_4x_5 - C_{l2}x_4 + C_{l3}x_6] \\ &-\frac{1}{\Delta}[-I_{xz}(C_{l5}\beta + C_{l1}\dot{\beta} + C_{l7}) + I_{xx}(C_{l5}\beta + C_{l1}\dot{\beta}) - C_{n7}] \end{aligned}$$

$$f_7(x) = x_4 + x_5\sin x_7 \tan x_8 + x_6\cos x_7 \tan x_8$$

$$f_8(x) = x_5\cos x_7 - x_6\sin x_7$$

$$f_9(x) = \frac{x_5\cos x_7 + x_6\sin x_7}{\cos x_8}$$

$$g(x) = \begin{bmatrix} \dfrac{F_{prop}\cos\alpha_m}{m} & C_{x3} & 0 & 0 \\ 0 & 0 & C_{y4} & C_{y5} \\ \dfrac{F_{prop}\sin\alpha_m}{m} & C_{z3} & 0 & 0 \\ 0 & 0 & a_1 & a_2 \\ 0 & C_{m3} & 0 & 0 \\ 0 & 0 & a_3 & a_4 \\ 0 & 0 & 0 & 0 \\ 0 & 0 & 0 & 0 \\ 0 & 0 & 0 & 0 \end{bmatrix} \tag{5}$$

where $\Delta = I_{xz}^2 - I_{xx}I_{zz}$, $a_1 = -\frac{(I_{zz}C_{l4} - I_{xz}C_{n4})}{\Delta}$, $a_2 = -\frac{(I_{zz}C_{l6} - I_{xz}C_{n5})}{\Delta}$, , $a_4 = -\frac{(I_{zz}C_{n5} - I_{xz}C_{l6})}{\Delta}$.

The coefficients C_{x1}, ..., C_{n5} are defined in **Table 3** [21, 22, 25, 26].

$C_{x1} = \frac{QSC_{x\dot{\alpha}}}{m}$	$C_{x2} = \frac{QScC_{xq}}{mV}$	$C_{x3} = \frac{QSC_{x\delta e}}{m}$	$C_{x4} = \frac{QSC_{x0}}{m}$
$C_{x5} = \frac{QSC_{x\alpha}}{m}$	$C_{y1} = \frac{QSbC_{y\dot{\beta}}}{2mV}$	$C_{y2} = \frac{QSbC_{yp}}{2mV}$	$C_{y3} = \frac{QSbC_{yr}}{2mV}$
$C_{y4} = \frac{QSC_{y\delta a}}{m}$	$C_{y5} = \frac{QSC_{y\delta r}}{m}$	$C_{y6} = \frac{QSC_{y\beta}}{m}$	$C_{y7} = \frac{QSC_{y0}}{m}$
$C_{z1} = \frac{QScC_{z\dot{\alpha}}}{mV}$	$C_{z2} = \frac{QScC_{zq}}{mV}$	$C_{z3} = \frac{QSC_{z\delta e}}{m}$	$C_{z4} = \frac{QSC_{z0}}{m}$
$C_{z5} = \frac{QSC_{z\alpha}}{m}$	$C_{l1} = \frac{QSc^2C_{l\dot{\beta}}}{2V}$	$C_{l2} = \frac{QSb^2C_{lp}}{2V}$	$C_{l3} = \frac{QSb^2C_{lr}}{2V}$
$C_{l4} = QSbC_{l\delta a}$	$C_{l5} = QSbC_{l\beta}$	$C_{l6} = QSbC_{l\delta r}$	$C_{l7} = QSbC_{l0}$
$C_{m1} = \frac{QSc^2C_{m\dot{\alpha}}}{2V}$	$C_{m2} = \frac{QSc^2C_{mq}}{V}$	$C_{m3} = \frac{QScC_{m\delta e}}{I_{yy}}$	$C_{m4} = QScC_{m0}$
$C_{m5} = QScC_{m\alpha}$	$C_{n1} = \frac{QSb^2C_{m\dot{\beta}}}{2V}$	$C_{n2} = \frac{QSb^2C_{np}}{2V}$	$C_{n3} = \frac{QSb^2C_{nr}}{2V}$
$C_{n4} = QSbC_{n\delta a}$	$C_{n5} = QSbC_{n\delta r}$	$C_{n6} = QSbC_{n\beta}$	$C_{n7} = QSbC_{n0}$

Table 3. Expression of the modified aerodynamic coefficients.

6. Integral sliding mode controller problem formulation

Consider the following nonlinear uncertain system [31]

$$\begin{aligned} \dot{x} &= f(x) + g(x).U \\ y &= S(x,t) \end{aligned} \tag{6}$$

$S(x,t)$ is a sliding variable. f and g are uncertain smooth vector fields and are differentiable.

The uncertainties in $f(x)$ and $g(x)$ are caused by the parameter variations, the nonmodeled dynamics, or the external disturbances.

Assumption 1 [31]: The relative degree r of system (6) is constant and known, and the associated zero dynamics are stable.

The rth-order sliding mode is defined through the following definition.

Definition 1 [6, 7, 8, 31]: Consider the nonlinear system (6) and the sliding variable S. Assume that the time derivatives $S, \dot{S}, \ldots.., S^{(r-1)}$ are continuous functions. The manifold defined as

$$\Sigma^r = \{x | S(x,t) = \dot{S}(x,t) = \ldots.. = S^{(r-1)} = 0\} \tag{7}$$

is called "rth-order sliding mode set," which is nonempty and is locally an integral set in the Fillipov sens [30]. The motion Σ^r on is called "rth-order sliding mode" with respect to the sliding variable S.

Definition 2 [6–8, 31, 32]: Consider the nonlinear system (6) and the sliding variable S. Assume that the time derivatives $S, \dot{S}, \ldots.., S^{(r-1)}$ are continuous functions. The manifold defined as

$$\Sigma_*^r = \{x | |S| \le \mu_0 \tau^{r-1}, |\dot{S}| \le \mu_1 \tau^{r-1}, \ldots.., |S^{(r-1)}| \le \mu_r\} \tag{8}$$

With $\mu_i \geq 0$ $(0 \leq i \leq r-1)$, is named "real rth-order sliding mode set," which is nonempty and is locally an integral set in the Fillipov sens [30]. The motion on Σ^r is called "real rth-order sliding mode" with respect to the sliding variable S. Given the form of system (6), the rth-order sliding mode control (SMC) approach allows the finite time stabilization to zero of the sliding variable S and its (r-1) first time derivatives by defining a suitable discontinuous control function. The rth time derivative of S satisfies the equation [6–8]:

$$S^{(r)} = a(x,t) + b(x,t)U \tag{9}$$

With $b = L_g L_f^{r-1} S$ and $a = L_f^r S$

Assumption 2 [31, 32]**:** Solutions of Eq. (9) with discontinuous right-hand side are defined in the sense of Fillipov [30].

Assumption 3 [31, 32]**:** Functions $a(t,x)$ and $b(t,x)$ are smooth and uncertain but bounded functions; furthermore, they can be partitioned into a well-known nominal part (respectively, $\bar{a}(t,x)$ and $\bar{b}(t,x)$ is an uncertain bounded one, respectively, $a(t,x)$ and $\Delta b(t,x)$.

$$\begin{aligned} a(t,x) &= \bar{a}(t,x) + \Delta a(t,x) \\ b(t,x) &= \bar{b}(t,x) + \Delta b(t,x) \end{aligned} \tag{10}$$

Functions $a(t,x)$ and $\bar{a}(t,x)$ are such that $a > 0$ and $\bar{a} > 0$ there is an upper bound constant ξ and a priori known constant $0 < \gamma \leq 1$ such that the uncertain functions satisfy the following inequalities [33]:

$$\left| \frac{\Delta b(t,x)}{\bar{b}(t,x)} \right| \leq 1 - \gamma, \; |\Delta a(t,x)| \leq \xi \tag{11}$$

The rth-order sliding mode controller (SMC) of Eq. (6) with respect to the sliding variable S is equivalent to the finite time stabilization of

$$\begin{aligned} \dot{z}_i &= z_{i-1} \\ \dot{z}_i &= a(t,x) + b(t,x) \end{aligned} \tag{12}$$

With $1 \leq i \leq r-1$ and $z = [z_1 \quad z_2 \quad \ldots \quad \ldots \quad z_r]^T = [S \quad \dot{S} \quad \ldots \quad \ldots. \quad S^{(r-1)}]^T$

Consider the following state feedback control

$$U = \frac{1}{\bar{b}(x,t)} \left(-\bar{a}(t,x) + \sigma \right) \tag{13}$$

with σ the auxiliary control input. Note that this state feedback control linearizes (by an input-output point of view) the nominal system, i.e., system (12) with no uncertainties.

Applying Eq. (13) to system (10), one gets

$$\dot{z}_i = z_{i-1}$$
$$\dot{z}_i = \Delta a(t,x) - \frac{\Delta b(t,x)}{\bar{b}(t,x)}\bar{a}(t,x) + \left(1 + \frac{\Delta b(t,x)}{\bar{b}(t,x)}\right)\sigma \tag{14}$$

The control objective is now the following: how to define a discontinuous control law ensuring the stabilization of the previous system, in a finite time and in spite of the uncertainties?

6.1. Control design

We proposed two high-order sliding mode controllers based on integral sliding mode concept [27]: the first requires knowledge of the uncertainties bounds, whereas, for the second one, no knowledge of the bounds is required. This latter feature is due to an adaptation law for the control gain.

6.1.1. Finite time stabilization of an integrators' chain system

The following theorem proposes a continuous finite time stabilizing feedback controller for a chain of integrators, by giving an explicit construction involving a small parameter. One gets an asymptotically stable closed-loop system; the system is homogeneous of negative degree with respect to a suitable dilation, which implies the finite time stability. Consider the system (12) with no uncertainty ($\Delta a(t,x) = 0$ and $\Delta b(t,x) = 0$).

$$\dot{z}_i = z_{i-1}$$
$$\dot{z}_r = \sigma \tag{15}$$

Theorem 1 [28]

Let $k_1, ..., k_r > 0$ be such that the polynomial $\lambda^r + k_r\lambda^{r-1} + ... + k_2\lambda + k_1$ is Hurwitz. There exists $\varepsilon \in]0,1[$ such that, for every $\alpha \in]1-\varepsilon, 1[$, the origin is a globally finite time stable equilibrium point for system (15) under the feedback

$$\sigma = k_1 sign(z_1)|z_1|^{\alpha_1} - \ldots - k_r sign(z_r)|z_r|^{\alpha_r} \tag{16}$$

With $\alpha_1, ..., \alpha_{r-1}$ satisfy $\alpha_{i-1} = \frac{\alpha_i \alpha_{i+1}}{2\alpha_{i+1} - \alpha_i}$

For $i = 2, ..., r$ with $\alpha_r = \alpha$ and $\alpha_{r+1} = 1$.

6.1.2. Robust finite time controller design based on integral sliding mode [31, 32]

Consider the following function, named "integral sliding variable," defined as (t_0 being the initial time)

$$S(z(t)) = z_r(t) - z_r(t_0) - \int_{t0}^{t} \sigma_{nom}(\tau)d\tau \tag{17}$$

with the term σ_{nom} defined by Eq. (16) in Theorem 1. Note that, $S(z(t_0)) = 0$: then the system is evolving on the sliding manifold early from the initial time.

This latter feature is a key point of the integral sliding mode controller; in fact, the definition of the integral sliding variable allows to ensure that a sliding mode has been established early from the initial time, thanks to the finite time convergence property of σ_{nom}. Then, it is necessary to force the system to evolve on the integral sliding surface $S = 0$ in spite of the uncertainties and perturbations: it will be the role of the discontinuous part of the controller. In fact, the term σ_{nom} appearing in S can be viewed as a desired trajectory generator. By supposing that, $\forall t \geq t_0, S = 0$, one has

$$\dot{S} = \dot{z}_r - \sigma_{nom} = 0 \rightarrow \dot{z}_r = \sigma_{nom} \tag{18}$$

From the previous inequality, it is clear that, if the control σ guarantees that $S = 0, \forall t \geq t_0$ and given the features of σ_{nom}, system (15) is stabilized at the origin in a finite time.

Then, in order to stabilize system (15), the following control law is defined

$$\sigma = \sigma_{nom} - Ksign(S) \tag{19}$$

This controller has two parts:

- The first one σ_{nom}, called "ideal control", is continuous and stabilizes the system (15) at the origin in absence of uncertainties. This controller is also used in order to generate the system's ideal trajectories;
- The second one $-Ksign(S)$ provides the complete compensation of uncertainties and perturbations and ensures that control objectives are reached, where the gain is satisfying

$$K > \frac{(1-\gamma)(|\sigma_{nom}| + |\psi| + \xi + \eta)}{\gamma} \tag{20}$$

Theorem 2: [29, 33] Consider the nonlinear system (6) and assume that assumptions 1–3 are fulfilled. Then, if the gain

K fulfills the condition (20), the control law

$$U = b^{-1}(x,t)\Big(-\bar{a}(x,t) + \sigma_{nom} - Ksign(S)\Big) \tag{21}$$

ensures the establishment of a rth-order sliding mode versus the sliding variable S, i.e., the trajectories of system (6) converge to zero in finite time.

7. Application of the adaptive integral- high-order-sliding -mode controller for piloting

The relative degrees are $r_\varphi = r_\theta = r_\psi = 0$.

The input control U is defined by $\sigma_{\varphi,\theta,\psi} = \begin{bmatrix} \delta_e & \delta_a & \delta_r \end{bmatrix}^T$.

http://dx.doi.org/10.5772/67580

We propose the integral sliding variable as follows:

$$S_{\varphi,\theta,\psi}(z(t)) = z_{1,\phi,\theta,\psi}(t) - y_d(t_0) - \int_{t0}^{t} \sigma_{nom}(\tau)d\tau \tag{22}$$

where $y_d = [\varphi_d \quad \theta_d \quad \psi_d]^T$ is the desired vector and $z_{1,\varphi,\theta,\psi} = [\varphi \quad \theta \quad \psi]^T$ is the output vector of integrators' chain.

In Theorem 1, we choose $\varepsilon = 0.7$, so we can take $\alpha = 0.5$.

The integrators' chain is defined by

$$\begin{aligned} \dot{z}_1 &= z_2 \\ \dot{z}_2 &= -\hat{\lambda}_{1\varphi,\theta,\psi}|z_{1\varphi,\theta,\psi}|^{\frac{1}{3}} sign(z_{1\varphi,\theta,\psi}) - \hat{\lambda}_{2\varphi,\theta,\psi}|z_{2\varphi,\theta,\psi}|^{\frac{1}{2}} sign(z_{2\varphi,\theta,\psi}) \end{aligned} \tag{23}$$

where, $\sigma_{nom} = -\hat{\lambda}_{1\varphi,\theta,\psi}|z_{1\varphi,\theta,\psi}|^{\frac{1}{3}} sign(z_{1\varphi,\theta,\psi}) - \hat{\lambda}_{2\varphi,\theta,\psi}|z_{2\varphi,\theta,\psi}|^{\frac{1}{2}} sign(z_{2\varphi,\theta,\psi})$.

The control input can be chosen as

$$\begin{aligned} \sigma_{\varphi,\theta,\psi} = &-\hat{\lambda}_{1\varphi,\theta,\psi}|z_{1\varphi,\theta,\psi}|^{\frac{1}{3}} sign(z_{1\varphi,\theta,\psi}) - \hat{\lambda}_{2\varphi,\theta,\psi}|z_{2\varphi,\theta,\psi}|^{\frac{1}{2}} sign(z_{2\varphi,\theta,\psi}) \\ &- \hat{\lambda}_{3\varphi,\theta,\psi}\int_0^t sign(z_{2\varphi,\theta,\psi})dt - K_{1\varphi,\theta,\psi} z_{2\varphi,\theta,\psi} \end{aligned} \tag{24}$$

where $K_{1\varphi,\theta,\psi} \succ 0$.

The reduction of the noise is assumed by the presence of the linear term ($K_i z_{2i}$, where $i = \varphi, \theta, \psi$) in the equation of each output i in the algorithm. This linear term can be expressed as the law of the control, which allows the reduction of the chattering effect. The addition of this continuous term smoothes the output noise due to a low gain values. If the chosen values of these gains become very low, the convergence of the algorithm becomes slow. Therefore, the choice of the convergence gains remains difficult and is based on a compromise between reducing the noise and having a short algorithm's convergence time. It should also be noted that in the presence of noise, it is necessary to impose small initial values for the dynamic gains in order to reduce the effect of the discontinuous control. Moreover, the presence of integral term ($\int_0^t sign(z_{2\varphi,\theta,\psi})dt$) in the expressions of the dynamic gains provides the smoothing of the estimated derivatives.

The dynamic adaptation of the gains $\dot{\hat{\lambda}}_i, i \in \{0,1,2\}$ is given by

$$\begin{cases} \dot{\hat{\lambda}}_{1\varphi,\theta,\psi} = |z_{1\varphi,\theta,\psi}|^{\frac{2}{3}} sign(z_{1,\varphi,\theta,\psi}) z_{1,\varphi,\theta,\psi} \\ \dot{\hat{\lambda}}_{2\varphi,\theta,\psi} = |z_{2\varphi,\theta,\psi}|^{\frac{1}{2}} sign(z_{2,\varphi,\theta,\psi}) z_{2,\varphi,\theta,\psi} \\ \dot{\hat{\lambda}}_{3\varphi,\theta,\psi} = z_{2\varphi,\theta,\psi}\int_0^t sign(z_{2,\varphi,\theta,\psi})dt \end{cases} \tag{25}$$

The application of this piloting technique in FS2004 is shown in **Figure 2**. λ, μ and h are latitude, longitude and altitude of aircraft, respectively.

The input signals at the upper and lower saturation values of the control laws are used to respect the actuators bounds. Scaled functions are added to take into account the actuators resolutions.

The adaptive integral high-order sliding mode technique is used to recover the desired signal. Several flight tests were realized to demonstrate the effectiveness of the combined controller/integrators' chain.

7.1. Simulation results

We run the flight simulator FS2004 and the interface with the module real-time windows target of Simulink/Matlab.

In a first step, we used aircraft predator, the aircraft taking off was done using the keyboard. Then, we run our software to transmit the control inputs based on the adaptive integral higher-order sliding mode to the autopilot controller in order to maintain the desired trajectory.

The desired signal injected and the output integrators' chain are shown in **Figure 4**. We notice the outputs of the integrators' chain $z_{1,j}$ where $j = \varphi, \theta, \psi$ follows the references φ_d, θ_d and ψ_d perfectly. The surface sliding mode $S_{\varphi,\theta,\psi}$ is small (see **Figure 5**).

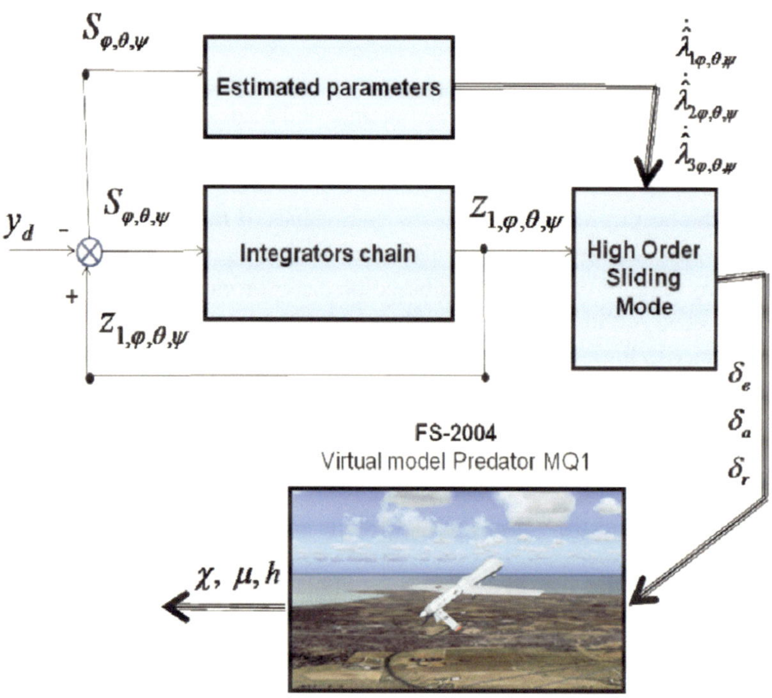

Figure 4. Application of the adaptive integral high order sliding mode controller in FS2004.

http://dx.doi.org/10.5772/67580

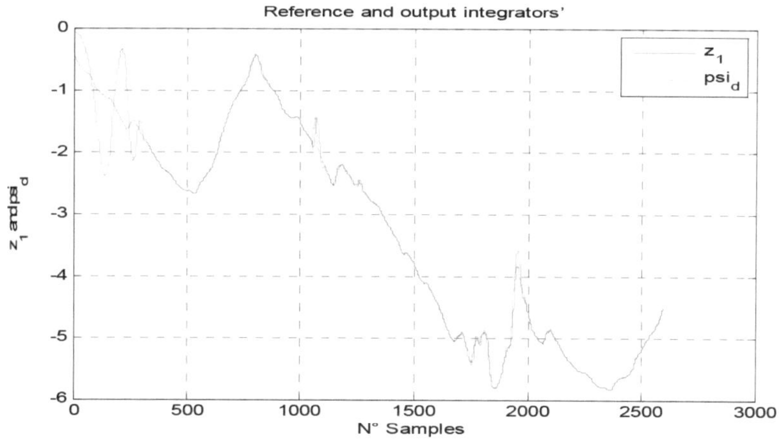

Figure 5. Reference and output integrators.

Figure 6 shows the error between the output integrators' chain $z_{1\varphi}$ and φ_{d0}. The signal $z_{1\varphi}$ follows φ_d.

The input signals at the upper and the lower saturation values of the aileron, rudder, and elevator deflections are used to respect the virtual Joystick (PPjoy) bounds. Upper limit: 62767, lower limit: 1.

Airwrench gives the following data:

- Aileron parameters: Aileron area 1.70 m^2, aileron up angle limit 20.0°, aileron down angle limit 15.0°.
- Elevator parameters: Elevator area 1.54 m^2, elevator up angle limit 25.00°, elevator down angle limit 20.00°.
- Rudder parameters: Rudder area 0.62 m^2, Rudder angle limit 24.00°.

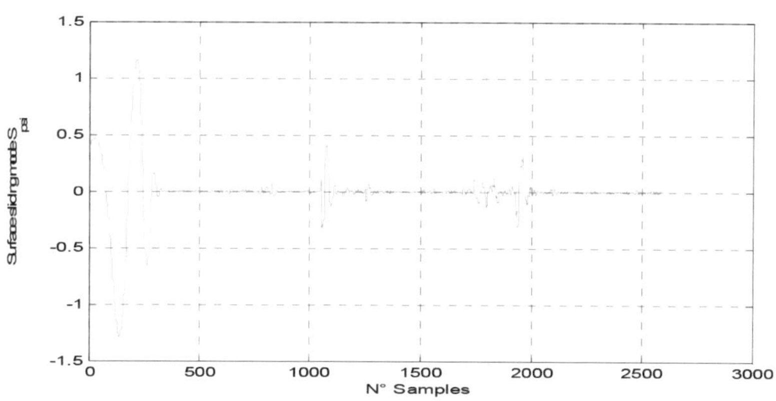

Figure 6. Surface sliding mode S_ϕ.

The aileron, elevator, and rudder deflections are shown in **Figures 7–9**. We notice the absence of the chattering phenomenon.

The evolution parameters $\hat{\lambda}_1$, $\hat{\lambda}_2$, and $\hat{\lambda}_3$ are shown in **Figure 10**.

The flight tests demonstrate the robustness of the adaptive integral high-order sliding mode. It makes it possible to ensure a better derivation of the desired input signal in real time, and this is to ensure a good accuracy of tracking the desired trajectory.

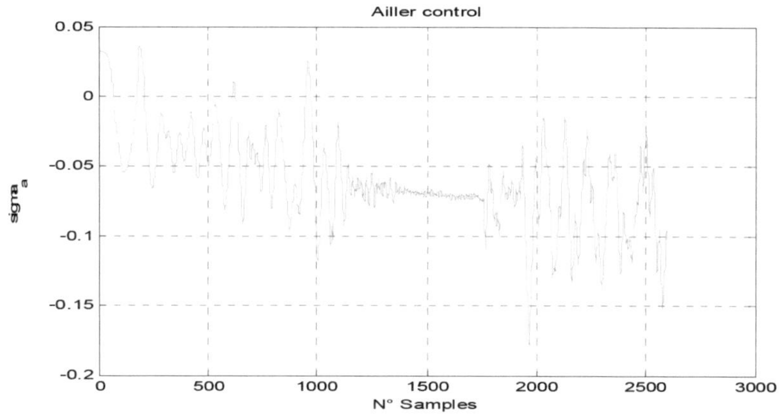

Figure 7. Ailler control.

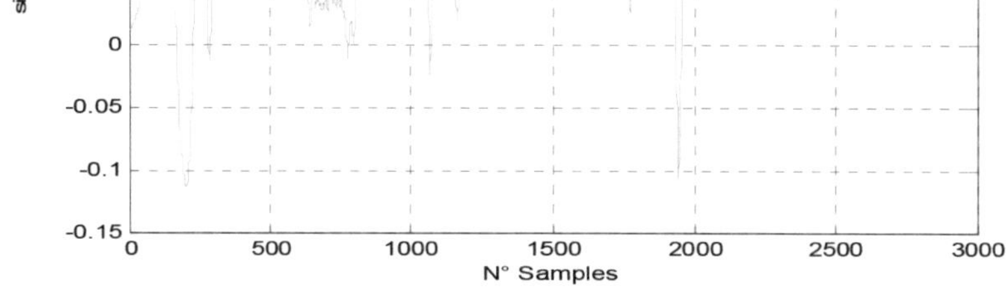

Figure 8. Rudder control.

http://dx.doi.org/10.5772/67580

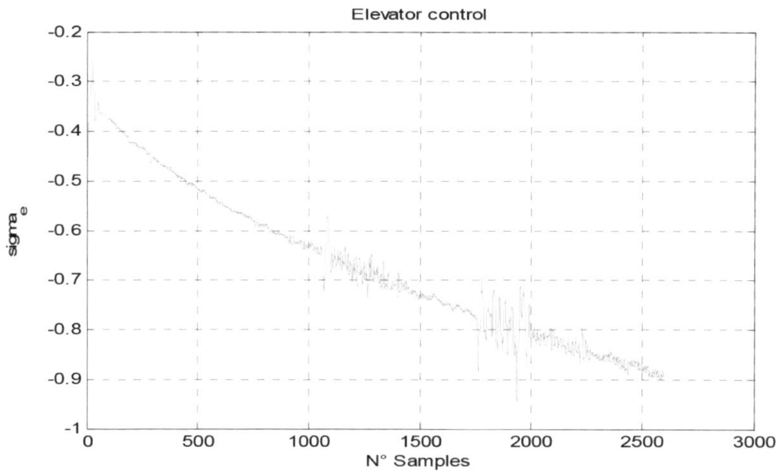

Figure 9. Elevator control.

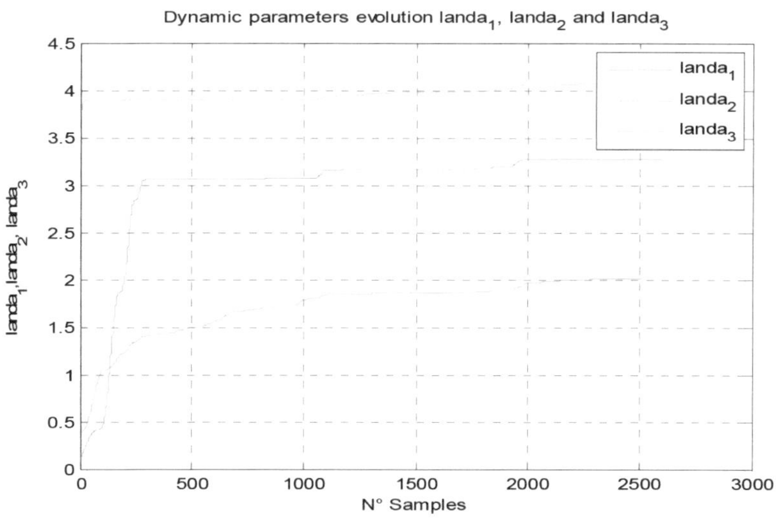

Figure 10. Dynamic parameters evolution $\hat{\lambda}_1$, $\hat{\lambda}_2$, and $\hat{\lambda}_3$.

8. Conclusion

In this chapter, a procedure of the communication with an aircraft model in a simulated environment and the implementation of the real-time interface between the Microsoft flight simulator and the module "real-time windows target" of Simulink/Matlab has been presented. After that, an adaptive integral sliding mode for an aircraft autopilot has been presented. Our approach uses the environment simulator (FS2004) to reduce the design process complexity.

For the piloting part, we have interested the gain adaptation for the reduction of chattering phenomena and possibility to control the aircraft presented by the uncertain nonlinear systems in which the uncertainties have unknown bounds. This technique is more robust and simpler to implement than the quaternion one and only needs the information about the sliding mode surface.

The flight tests demonstrate the robustness of an adaptive integral sliding mode. The former ensures a better derivation of the desired input signal in real time, and this ensures a good accuracy in terms of tracking for a desired reference.

Author details

Zaouche Mohammed*, Foughali Khaled and Amini Mohamed

*Address all correspondence to: zaouchemohamed@yahoo.fr

Control and Automation Laboratory, Ecole Supérieure Ali Chabati, Réghaia, Alger, Algeria

References

[1] O. Harkegard and S. Torkel Glad Flight Control Design Using Backstepping, Linkopings universitet, Linkoping, Sweden, 2001.

[2] J.J.E. Slotine and Li, Applied nonlinear control, Practice-Hall, Englewood Cliffs, New Jersey 07632, United States, 1991

[3] J.J.E. Slotine and J.A. Coetsee, Adaptive sliding controller synthesis for non-linear systems, International Journal of Control, Vol.43, Issue 6, pp. 1631–1651, 1986.

[4] J.L. Junkins, K. Subbarao and A. Verma, Structured adaptive control for poorly modeled nonlinear dynamical systems. Computer Modeling in Engineering & Sciences, Vol. 1, No. 4, pp. 99–118, 2000.

[5] V. Chiroi, L. Munteanu and I. Ursu, On chaos control in uncertain nonlinear system. Computer Modeling in Engineering & Sciences, Vol. 72, No. 3, pp. 229–246, 2011.

[6] A. Levant, Higher-order sliding modes, differentiation and output feedback control. International Journal of Control, Vol. 76, No. 9/10, pp. 924–941, 2003.

[7] A. Levant, Robust exact differentiation via sliding mode technique. Automatica, Vol. 34, No. 3, pp.379–384, 1989.

[8] A. Levant, Quasi continuous high order sliding mode controllers, IEEE Transactions on Automatic Control, Vol.50, No. 11, pp. 1812–1816, 2005

[9] A. Sabanovic, L. Fridman, S. Spurgeon, Variable structure systems: from principles to implementation, The Institution of Engineering and Technology, ISBN: 978-0-86341-350-6, 2004

[10] B. Bandyopadhyay and J. Sivaramakrishnan, Discrete-time Sliding Mode Control: A Multirate Output Feedback Approach, Lecture Notes in Control and Information Sciences, Springer Berlin Heidelberg New York, ISBN 978-3-540-28140-5, pp. 27–49, 2006

[11] B. Bandyopadhyay, F. Deepak and K.-S. Kim, Sliding Mode Control Using Novel Sliding Surfaces, Lecture Notes in Control and Information Sciences, Vol. 392, Springer Berlin Heidelberg New York. ISBN 978-3-642-03448-0, 2009

[12] D. Allerton, Principles of flight simulation, Aerospace series, John Wiley & Sons, Ltd. ISBN: 978-0-470-75436-8, pp. 97–154, 2009

[13] G. Bartolini, L. Fridman, A. Pisano and E. Usai, Modern Sliding Mode Control Theory: New Perspectives and Applications, Lecture Notes in Control and Information Sciences, Vol. 375, Springer, Berlin Heidelberg New York. ISBN 978-3-540-79016-7, chap. 4, 2008

[14] H. Yigeng, S. Laghrouche, Liu Weiguo and A. Miraoui, Robust High Order Sliding Mode Control of Permanent Magnet Synchronous Motors, Book: Recent Advances in Robust Control - Theory and Applications in Robotics and Electromechanics, InTech. Chapter 13, ISBN 978-953-307-421-4, 2011

[15] W. Perruquetti and J.P. Barbot, Sliding mode control in engineering, Marcel Dekker, New york Basel, ISBN:0-8247-0671-4, 2000

[16] J.L. Boiffier, The dynamics of flight: the equations, John Wiley & Sons, Chichester UK, ISBN: 0471942375, pp. 92, 1998

[17] P. Lopez and A. S. Nouri, Elementary and practical theory of the sliding mode controls, Springer Berlin Heidelberg New York. ISBN 3-540-31003-7, 2000

[18] R. Louali, R. Belloula, M.S. Djouadi and S. Bouaziz, Real-time characterization of Microsoft Flight Simulator 2004 for integration into Hardware in the Loop architecture, 19th Mediterranean Conference on Control and Automation, Greece, 2004.

[19] J. Salgado, Contribution to the control of an autonomous submarin robot sous marin type torpedo, Phd Thesis, Université de Montpellier, 2004.

[20] V.I. Utkin, Sliding Mode in Control Optimisation, Springer-Verlag, Berlin, 1992.

[21] V. Klein, P.C. Murphy, T.J. Curry, and J.M. Brandon, Analysis of Wind Tunnel Longitudinal Static and Oscillatory Data of the F-16XL Aircraft, NASA/TM-97-206276, December 1997.

[22] Jay M. Brandon and John V. Foster, Recent dynamic measurements and considerations for aerodynamic modeling of fighter airplane configurations, American Institute of Aeronautics and Astronautics, AIAA – 98–4447, 1998.

[23] Yuri B. Shtessel. Ilya A. Shkolnikov, Mark, D J. Brown An asymptotic second-order smooth sliding mode control, Asian Journal of Control. Vol. 5, No. 4, pp. 498–504, December 2003.

[24] W. Perruquetti and J. P. Barbot, Sliding mode control in engineering, Marcel Dekker, New York. 2002.

[25] M. Zaouche, Identification and robust control of an aerodynamic system with three axes, Phd Thesis, School Military Polytechnic, Algiers, 2015.

[26] M. Zaouche, A. Beloula, R. louali, S. Bouaziz and M. Hamerlain Adaptive differentiators via second order sliding mode for a fixed wing aircraft, Computer Modeling in Engineering and Sciences, Vol. 104, No. 3, pp. 159–184, 2015.

[27] S. Laghrouche, F. Plestan, A. Glumineau, Higher order sliding mode control based on integral sliding mode Automatica,Vol. 43, No. 3, pp.531–537, 2007.

[28] M. Defoort, T. Floquet and A. Kokosy, Finite time control of a class of MIMO nonlinear systems using high order integral sliding mode control Proc. 9th Int. Conf. on Variable Structure Systems, Alghero, Italy, pp. 133–138, 2006.

[29] Q. Zong, , Z. S Zhao and J. Zhang Higher order sliding mode control with self-tuning law based on integral sliding mode, IET Control Theory and Application, Vol. 4, No. 7, pp. 1282–1289, 2010.

[30] A.F. Filippov, Differential equations with discontinuous right hand side, Kluwer Academic Publisher, Dordrecht, 1988.

[31] M. Taleb, , F. Plestan and B. Bououlid. An adaptive solution for robust control based on integral high-order sliding mode concept: adaptive integral sliding mode control, International Journal of Robust and Nonlinear Control, 2014.

[32] J. Zhang, Q. Zong, and Z.-S. Zhao. Higher order sliding mode control with self-tuning law based on integral sliding mode, IET Control Theory and Applications, Vol. 4, No. 7, pp. 1282–1289, 2010.

[33] Qun Zong, Zhanshan Zhao and Liqian Dou, Higher order adaptive sliding mode control for a class of SISO systems, Proceedings of the 48h IEEE Conference on Decision and Control (CDC) held jointly with 2009 28th Chinese Control Conference, 28th Chinese Control Conference. Shanghai, 12/2009

[34] F. Plestan. A new algorithm for high-order sliding mode control, International Journal of Robust and Nonlinear Control, John Wiley & Sons, Ltd., Volume 18, Issue 4–5, pp. 441–453, 2008.